The Intelligent Enterprise

Contributions to Management Science

H. Dyckhoff/U. Finke
Cutting and Packing in Production and Distribution
1992. ISBN 3-7908-0630-7

R. Flavell (Ed.)
Modelling Reality and Personal Modelling
1993. ISBN 3-7908-0682-X

M. Hofmann/M. List (Eds.)
Psychoanalysis and Management
1994. ISBN 3-7908-0795-8

R.L. D'Ecclesia/S.A. Zenios (Eds.)
Operations Research Models in Quantitative Finance
1994. ISBN 3-7908-0803-2

M.S. Catalani/G.F. Clerico
Decision Making Structures
1996. ISBN 3-7908-0895-4

M. Bertocchi/E. Cavalli/S. Komlósi (Eds.)
Modelling Techniques for Financial Markets and Bank Management
1996. ISBN 3-7908-0928-4

H. Herbst
Business Rule-Oriented Conceptual Modeling
1997. ISBN 3-7908-1004-5

C. Zopounidis (Ed.)
New Operational Approaches for Financial Modelling
1997. ISBN 3-7908-1043-6

K. Zwerina
Discrete Choice Experiments in Marketing
1997. ISBN 3-7908-1045-2

G. Marseguerra
Corporate Financial Decisions and Market Value
1998. ISBN 3-7908-1047-9

WHU Koblenz – Otto Beisheim Graduate School of Management (Ed.)
Structure and Dynamics of the German Mittelstand
1999. ISBN 3-7908-1165-3

A. Scholl
Balancing and Sequencing of Assembly Lines
1999. ISBN 3-7908-1180-7

E. Canestrelli (Ed.)
Current Topics in Quantitative Finance
1999. ISBN 3-7908-1231-5

W. Bühler/H. Hax/R. Schmidt (Eds.)
Empirical Research on the German Capital Market
1999. ISBN 3-7908-1193-9

M. Bonilla/T. Casasus/R. Sala (Eds.)
Financial Modelling
2000. ISBN 3-7908-1282-X

S. Sulzmaier
Consumer-Oriented Business Design
2001. ISBN 3-7908-1366-4

C. Zopounidis (Ed.)
New Trends in Banking Management
2002. ISBN 3-7908-1488-1

U. Dorndorf
Project Scheduling with Time Windows
2002. ISBN 3-7908-1516-0

B. Rapp/P. Jackson (Eds.)
Organisation and Work Beyond 2000
2003. ISBN 3-7908-1528-4

M. Grossmann
Entrepreneurship in Biotechnology
2003. ISBN 3-7908-0033-3

H.M. Arnold
Technology Shocks
2003. ISBN 3-7908-0051-1

T. Ihde
Dynamic Alliance Auctions
2004. ISBN 3-7908-0098-8

J. Windsperger/G. Cliquet/G. Hendrikse/M. Tuunanen (Eds.)
Economics and Management of Franchising Networks
2004. ISBN 3-7908-0202-6

K. Jennewein
Intellectual Property Management
2004. ISBN 3-7908-0280-8

Markus J. Thannhuber

The Intelligent Enterprise

Theoretical Concepts and Practical Implications

With 32 Figures
and 3 Tables

Physica-Verlag
A Springer Company

Series Editors
Werner A. Müller
Martina Bihn

Author

Dr. Markus J. Thannhuber
Hans Einhell AG
Wiesenweg 22
94405 Landau an der Isar
Germany
thannhuberm@einhell.de

ISSN 1431-1941
ISBN 3-7908-1555-1 Physica-Verlag Heidelberg New York

Cataloging-in-Publication Data applied for
Library of Congress Control Number: 2004111904

Physica-Verlag is a part of Springer Science+Business Media

springeronline.com

Printed in Germany

Softcover design: Erich Kirchner, Heidelberg

SPIN 11015642   88/3130/DK-5 4 3 2 1 0 – Printed on acid-free paper

May this work be a contribution to the scientific community,
may it point in the right direction, provide clarity and not confusion
and may it stimulate our critical perception.

Preface

In today's competitive environments enterprises face diminishing market life spans, increasing pressure on profit margins and increasingly complex customer requirements. Thus in their operations, modern organizations have to find a high-level balance between dynamics, complexity and precision in order to best utilize their markets. Organization Theory and Industrial Engineering, the disciplines on hand helping industry to cope with this challenge, soon identified process optimizations as the key to possible solutions. Many efforts have been undertaken to provide sound theoretical models to deal with complexity and dynamics and streamline business processes. These efforts on the one hand helped companies to be more precise in carrying out their actions and even provided solutions to produce customized products at near-mass production prices (Mass-Customization). On the other hand it turned out to be one of the most difficult tasks to generalize and transfer experiences gained in one process-reengineering project to another and put the theoretical models into practice. Not without reason is it the extremely high failure rate of business-process-reengineering projects that today deters most enterprises from entering such adventures.

Right at the same time there emerged a new and highly promising scientific branch, Knowledge Management, that attracted many disciplines – among others again Organization Theory and Industrial Engineering. Knowledge was identified as a major production factor. In industrialized countries, value added is mainly raised by the intellectual abilities of a company's workforce. Knowledge Management became a leading topic for researchers and business consultants all over the world. Knowledge Management Models where developed, infrastructure built up and personal, professional insights, techniques and methodologies communicated, gathered and redistributed. The newly emerged learning culture but also IT tools and a newly gained focus on intangible assets reportedly helped numerous companies to speed-up product development, reduce costs and strengthen their overall organizational efficiency. However does it seem difficult to preserve these benefits over time so that many promising projects at the end did fall short of their expectations. And again it seems difficult to abstract and transfer experiences and insights from one project to another.

Only recently researchers suspected a certain relation between organizational knowledge and business processes: the employees' knowledge might manifest in the organizations' processes. But in fact they are the two sides of the medal – without scientifically understanding the one it is difficult to grasp the other. Knowledge on an organizational level is rooted deeply in the network of organizational processes and vice versa. Dynamic and adaptive enterprise processes can only be realized through organizational knowledge, precision only be achieved in the context of an intelligent organizational framework and complexity only be handled through the interplay of knowledge and intelligence. Today's approaches in most cases are merely a conceptualization of activities pragmatically organizing and implementing the efforts undertaken. There hardly exists a sound scientific basis or a theoretic framework – which makes it difficult to transfer insights from one case to another. Questions concerning the phenomenological nature of knowledge and intelligence on the enterprise level, how they emerge or how their development is supported and promoted are left unanswered. The first goal thus is to draft a theoretical model. Integrating perspectives from multiple disciplines – among which System Theory played a dominant role – Markus Thannhuber provided precise and to a certain extent new definitions for terms like the 'System Enterprise', 'Process', 'Declarative Processing' or 'Process Building Block' and introduced 'Knowledge' and 'Intelligence' as well defined concepts to dynamically generate tailor-made enterprise processes in response to a given set of stimuli. In a second step Markus Thannhuber identified structural and organizational consequences as well as practical implications for the modern enterprise, it's workforce and its processes. Last but not least should the theoretical model worked out as well as the resulting conceptual framework proof to be a feasible fundament for the implementation of practical tools and the necessary IT infrastructure.

Based on a theoretical foundation Markus Thannhuber discusses organization theoretic consequences and practical implications. The nature of control procedures and process building blocks leading to 'Declarative Processing' is described thoroughly and illustrated in a software prototype. The link between the ability of declarative processing, as a rather technical skill, and the emergence of cognitive phenomena on enterprise level is worked out and demonstrated by examples.

The transition towards intelligent enterprises requires drastic organizational changes. The 'Process Driven Management' has to be replaced by 'Project Driven Management', the employees have to be integrated as 'Knowledge Workers'. The role of the employee as basic source for 'Value Added' and in particular the role of the employees' capabilities as the interfaces between system 'Enterprise' and system 'Employee' are thoroughly discussed. Last but not least the need for a continuous evaluation is emphasized and suitable indicators are introduced to measure the degree to which organizational changes towards the intelligent enterprise are implemented and bear fruits.

With the presented work Markus Thannhuber introduces a theoretic framework and opens up new perspectives for Managerial Sciences, Organization Theory and Industrial Engineering. The resulting approach towards the intelligent enterprise moves the capability of dynamic processing and the structural and organizational framework to the center of interest. This indicates a sharp turn away from a number of traditional concepts such as the Learning Organization or traditional Knowledge Management. At the same time it demands for new scientific efforts in Industrial Engineering, the discipline that by tradition identifies the design, implementation and optimization of enterprise processes.

Hans-Jörg Bullinger
President of the Fraunhofer-Gesellschaft

Munich, July 2004

Acknowledgments

Throughout my work on this interesting subject I received invaluable support from many remarkable teachers and outstanding friends. In particular, I would like to thank Prof. Bullinger, who realized and supported the whole project and scientifically supervised this work. My special thanks go to Prof. Tseng and his family, who hosted this project for the biggest part. During my time at his institute I got to know him as an excellent professor, who strongly engaged in the project and contributed to the research. But I also got to know him as a personal advisor, teaching me far beyond the scope of this project. I would also like to thank Prof. Warschat and Dr. Fischer, who did an excellent job in supervising my research work. Of great importance to the success of this work was the contribution of all my fellow researchers, with whom I shared many interesting discussions. Therefore my special thanks go to Mr. Michael Chen, Mr. Agustino Kurniawan, Mr. Leslie Wong and Mrs. Merry Zhang. Not to forget all the fruitful discussions I had with Mr. Cornelius Kratzer and Mr. Johannes Beer, many thanks to them. I would also like to express my thanks to Mr. Robert Vilsmeier and Mr. Christian Wagner, who provided the necessary linguistic support to fine-tune this work.

As there is no harvest without a seed, I would also like to thank all those supporting the decisions towards my dissertation. Here my special thanks go to Prof. Becker, whom I trusted as a long-term personal advisor, and Dr. Arndt, who strongly encouraged me to take this step.

Finally, I would like to express my deepest thanks to my parents, Ms. Gisela Thannhuber and Mr. Josef Thannhuber, and to my brother Philipp Thannhuber, who strongly supported my education and gave me advice and help in any situation. And above all I would like to thank my fiancée Barbara Wimmer, who gave me the strength to carry out this research even though she had to abandon many of those things that were good and important to her.

Acknowledgements

Table of Contents

Chapter 1 Introduction

1.1 The Object of Discussion

The majority of today's companies thrives and prospers in economies that expose them to the harsh conditions of more or less liberalized markets. These economies allow businesses to move into the markets and serve unsatisfied needs, just as they have to accept businesses withdrawing from a market in threat of expected losses. The consumer benefits from a wide range of products to choose from, providing a high variety of features and a number of direct and indirect substitutes. In principle these markets still offer a fair chance of profit to all suppliers. To a certain extent all competitors have what is generally known as 'Market Power': the freedom to set the prices and manipulate the sales volume of their products.

However, these markets are no perfect markets but are what economists would call 'Markets with Monopolistic Competition' (MANKIW, 1998 [56]). Economic theories show that on the one hand the suppliers suffer, as they will never be able to produce at their efficient size of operation and thus underutilize their capacities (-> 'Excess Capacities'). The consumers on the other hand have to accept a net welfare loss, as the price they have to pay for any product will in general exceed the marginal costs of its production.

Companies, competing in a monopolistic competition[1], having the freedom of decision but still being bound to the market forces and ever striving for more and more customers to utilize their excess capacities shall be the object of our discussions.

Despite all the criticism Karl Marx raised against the subordination to the imperative of profit-making still every company strives to achieve a substantial income through its operation. Not only is profit a strong demand posed by the owners but it is also a necessary requirement for the existence and growth of an enterprise engaging in liberal markets of consumption and factors. Neglecting short-term distractions, as caused by speculative fantasies,

---

1 Although most of the effects described here could be easily transferred to an oligopoly, the discussion should be limited to the monopolistic competition.

emotional or group psychological effects, it is the enterprise's profit which is the only rationale behind its ability to attract and generate capital. With capital being one of the fundamental production factors – one of the essential input factors facilitating any operation – it is obvious that an enterprise not striving for profits would indeed lack a necessary requirement for its existence. In this context it should not surprise that profitability was stipulated to be the major yardstick for the validation of any activities by the management theories (DRUCKER, 1993 [24]).

1.2 Environmental Demands and Constraints

Operating in a market of monopolistic competition, it is this market's natural rules in force that will define the general framework for the company's operations and that will set the boundaries of its success. The economic theories describe the development of a market from its initial state, where a single company serves the whole market demand in a purely monopolistic fashion, to its final equilibrium, where a large number of suppliers compete for the limited expenditures that the consumers are willing to invest (MANKIW, 1998 [56]).

Taking first order approximations to the underlying cost and demand functions, the situation is best discussed with the help of 'Price/Cost'-'Quantity'-Diagrams (s. Figs. 1.1 a–c).

A company serving a new market as a sole player has almost any freedom to adjust its product price and produce for the market's demand. Thinking in 'marginal terms'[2] the company will soon find the optimal sales volume Q_{opt} exactly where the marginal revenue R_M equals the marginal costs C_M. With the price P_{opt} and the average costs per piece C_{Av}^{opt} both at Q_{opt}, the maximum profit is given by:

$$p_{\max} = Q_{opt}(P_{opt} - C_{Av}^{opt})$$

[2] Thinking in 'marginal terms' is a concept classically introduced by economists to reflect their theorem that the optimization of profits is the fundamental objective of every business operation. It should be noted that this assumption is not anymore considered to be true in a strong sense by management theorists. According to Drucker, the business purpose is to "create a customer" (DRUCKER, 1985 p. 61 [26]) and a market by means of marketing and innovation. Thus profit is not the rationale behind business activities but the only rational criterion for testifying their validity. As a consequence, businesses observed from 'outside' still appear to follow the maxim of profit maximization and thinking in 'marginal terms' – which should be sufficient for the purpose of the discussion in this work.

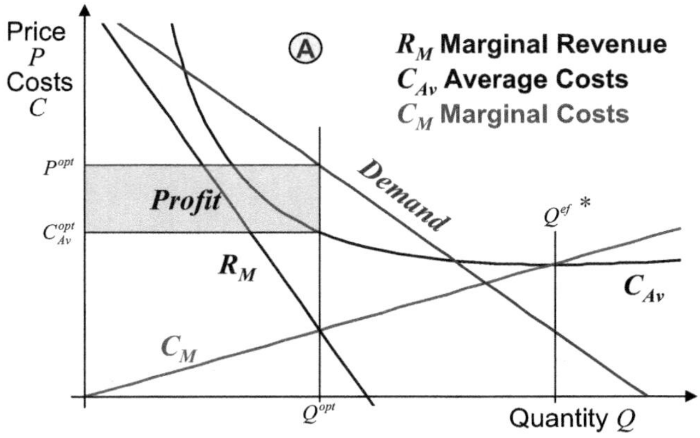

a) Initial Monopolistic Situation

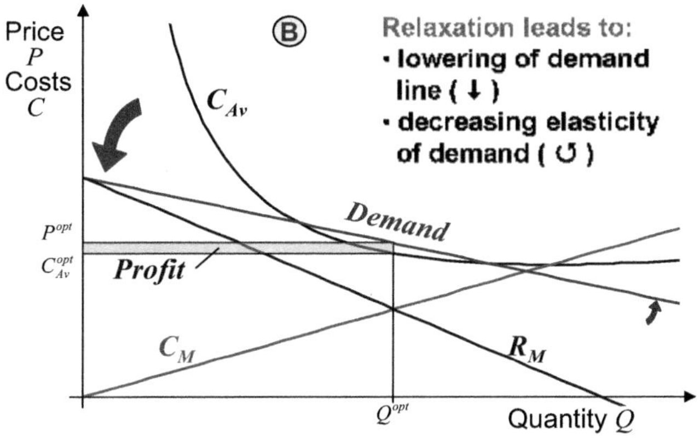

b) Market Relaxation

Fig. 1.1a,b. Companies in a monopolistic competition

The situation changes as competitors enter the market, offering similar products or even close substitutes. Now prices can only be varied within a narrow band as customers quickly decide to invest in alternatives (-> lowering of demand line and reduction of its elasticity). Companies that correctly adjust their parameters nevertheless should still be able to run profitably. A smaller profit, however, has to be realized by handling a significantly higher volume. (s. Fig. 1.1b).

Relaxation will continue to further lower and tilt the demand line until the market has reached its final equilibrium. At this stage all competitors have lost their 'Market Power' as they can sell their products only at one designated price P_{opt} without suffering losses. Even at this optimum the average

* Q^{es} The efficient size of operation (s. Glossary)

costs per piece C_{Av}^{opt} will count up exactly for the price P_{opt}, leaving no margin for profits (s. Fig. 1.1c). Now market entries and exits will slowly come to an end as there is no motivation to enter the market, but suppliers already in the market will still try to best utilize capacities built up and to pay for fixed costs. Relaxation has come to an end and the final equilibrium is reached.

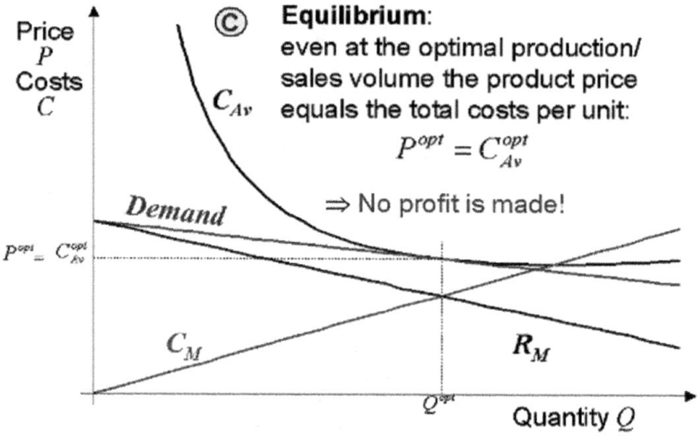

Fig. 1.1c. Companies in a monopolistic competition

Assuming that the motivation for entries and exits is directly proportional to the profit achievable at a given time *t*, it is easy to show that the temporal development of a market in the easiest case resembles an exponential decay over time (s. Fig. 1.1d).

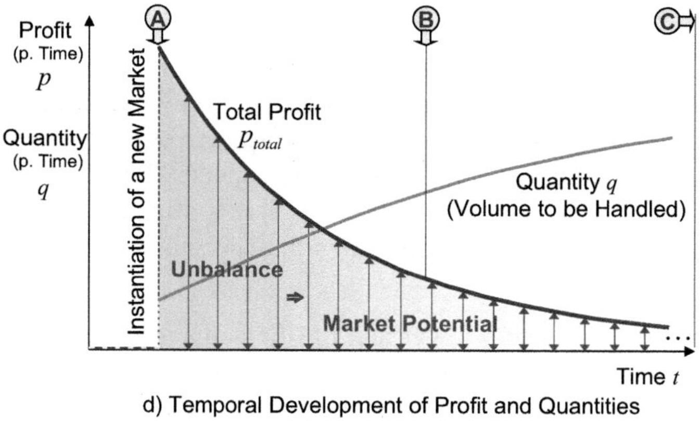

Fig. 1.1d. Companies in a monopolistic competition

Market unbalances are a vital necessity for every company. They are the potential differences exploited to cater for the energetic needs of any business operation. More precisely, it is the irreversible flow of products and its monetary counter flow, slowly driving the market's entropy S to its maximum, which turns the "Engine Enterprise", allowing it to convert some of the flow into assets. Being bound to the imperative of profit-making, every company has to find access to suitable markets offering substantial unbalances to exploit.

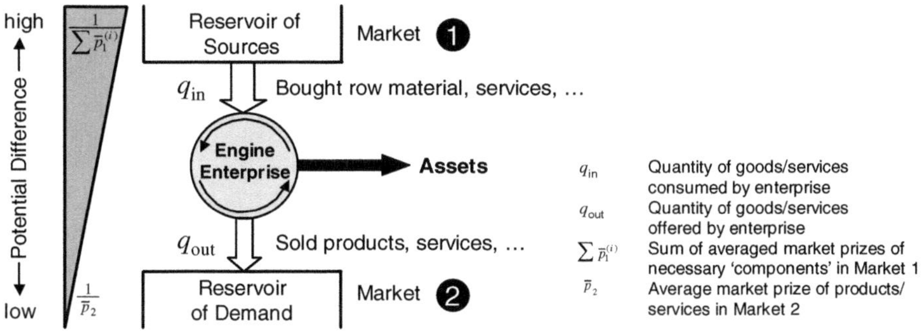

Fig. 1.2. The "engine" enterprise

1.3 Three Dominating Effects and the Necessities Implied

1.3.1 Dynamics

The relaxation of market unbalances induces inherent dynamics to every business. Have the timescales of relaxation been as long as several decades during the era of early industrialization, so do businesses in industries such as computer and computer components supply have to face average market life-cycles of several months and less. Major achievements in the information and communication technologies, advanced flexible production facilities and the globalization of trade caused timescales to tumble and will cause them tumbling even further in the future. As a consequence, every enterprise has to cope with an increasingly dynamic environment. Instead of just living on their niche they have to find new niches to exploit day by day. The ability to adapt effectively to the constantly changing environment is a key factor for survival.

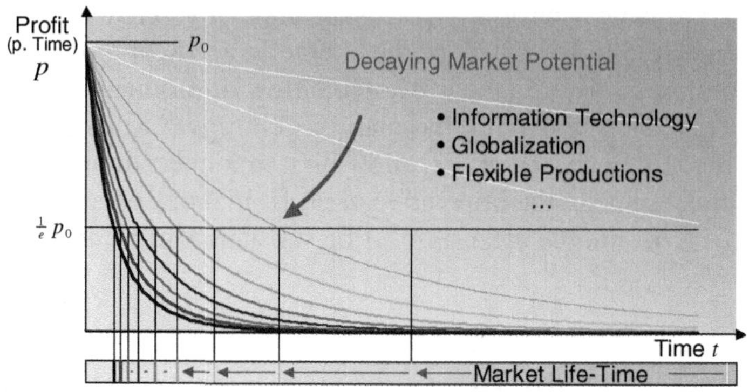

Fig. 1.3. Decaying market potentials

1.3.2 Precision

In the context of a market in monopolistic competition the existential demand for profit has a second important consequence: the need for highest precision. Fig. 1.1b vividly illustrates that a company has to precisely adjust its parameters to stay within the narrow band of profitability. Companies wrongly adjusting their price parameters or, even worse, companies that do not operate at or below the industries cost standards are quickly driven into losses. Fig. 1.1d clearly implies that precision gets even more important towards the end of the market life cycle. Due to the fact that businesses have to handle significantly higher volumes in order to exploit the diminishing market potential any arbitrary small variation in its variable costs causes a substantial threat of losses. With diminishing timescales quickly driving a market towards the lower part of the slope in the 'Profit'-'Time' diagram, precision today is at least equally important as adaptability.

1.3.3 Complexity

For any imaginable system to be able to dynamically adapt to a changing environment it requires sensors, actuators as well as the necessary degrees of freedom: sensors to measure and evaluate the status of the environment and itself, actuators to adjust the parameters that control the available degrees of freedom. Adaptation is the effort to evaluate the sensors' readings and adjust the parameters constantly to their 'near-to-optimum' positions. The degree of adaptation and with it the optimization achievable on the one hand depends on the precision of sensors, actuators and the underlying model that maps sensor readings to actions, but on the other hand it also depends on how well the environmental changes can be mapped at all to the available degrees of freedom. Once the parameters are fine-tuned to their optimum, the system needs to undergo revolutional changes to find access to new degrees of freedom in order to improve – a phenomenon that leads to what is widely known as the "Jumping S-Curves of Innovation" (FOSTER, 1986 [29]).

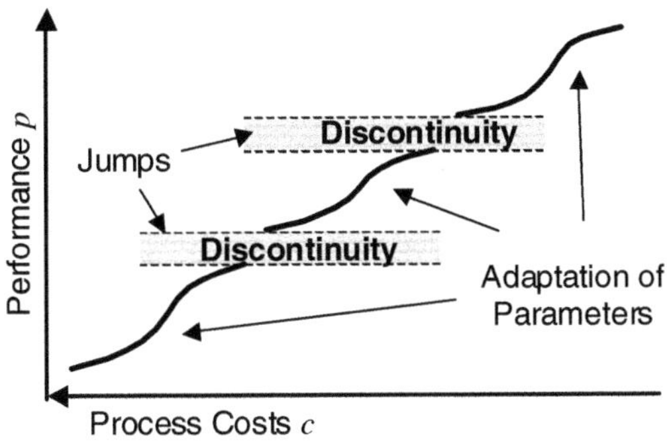

Fig. 1.4. The jumping S-curves

Transferred to the system enterprise this implies that every enterprise not only needs to have the sensors and actuators available to control its accessible degrees of freedom, but this also implies that it needs to be able to realize innovative jumps by finding access to new degrees of freedom whenever this is indicated. Of course, even though in a smaller scale, in reality every company does this day by day when applying new tools, new machines, new software packages or new skills that allow them to behave in a highly differentiated manner. But, every freedom gained is bought by an increasing complexity: it increases the amount of information to be processed, the number of actuators to be controlled and, last but not least, complicates any underlying decision models needed to derive effective behavior. The modern enterprise in every situation faces an enormous amount of possible next steps and needs to perceive a large number of stimuli to support their management decisions. Effective behavior relies on successfully mapping perceived patterns of stimuli to the right sets of actions. This requires a number of nontrivial and highly sensitive mappings.

The ability to deal with highest complexity, an enormous amount of information and almost unlimited alternatives to every step to take is a third necessary requirement for every enterprise to compete successfully.

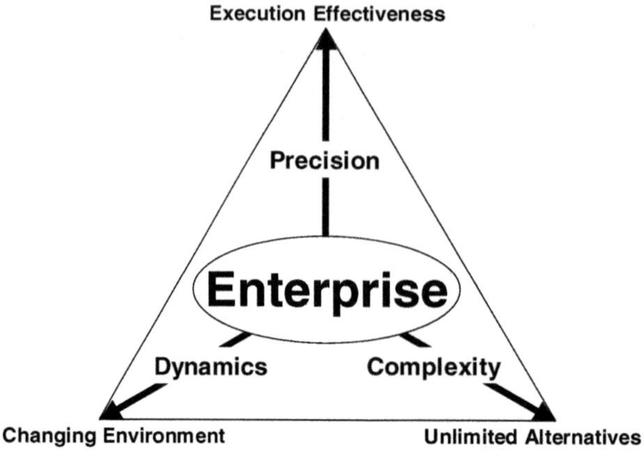

Fig. 1.5. The field of external forces

1.3.4 Conclusion

'Dynamics', 'Complexity' and 'Precision' are three highly coupled, interdependent effects that all have to be addressed simultaneously by any business operation. They represent the field of forces in which every company is torn. Each force pulling in the opposite direction a company has to struggle in order to survive, thrive and prosper.

By their very nature 'Dynamics', 'Complexity' and 'Precision' drive modern enterprises into a very challenging dilemma: mastering complexity is highly demanding in a dynamic environment, staying flexible and adaptive is difficult if precision is required and being precise is in particular challenging in a dynamic and complex environment. (THANNHUBER, TSENG and BULLINGER, 2001 [109])

1.4 Consequences for the Engineering Domain

With the manufacturing industry relying on their design, production and logistic processes, which are the dominant enablers of their operations, the engineering domain is inevitably exposed to the same challenging dilemma. Being a part of the whole, engineers organizing the product design process, planning the manufacturing process or arranging the logistic flow have to be able to address dynamics and complexity in their solutions while they still have to achieve the necessary precision in execution effectiveness.

If explicitly conscious about it or just implicitly aware the engineering community has well perceived this problem situation and brought up numerous domain-specific approaches to tackle several facets of this challenge. As early as 1985, Hatvany (HATVANY, 1985 [35]) pointed out that the rigidity of traditional hierarchical structures limited the dynamic performance of systems. Attempts for new approaches include O'Hare and Chisholm's proposed distributed artificial intelligence system (O'HARE and CHISHOLM, 1990 [68]) as well as advanced control methods such as predictive controllers introduced by Pritschow and Wiendahl (PRITSCHOW and WIENDAHL, 1995 [76]). The industry itself has adopted modern approaches like 'Mass Customization' to address the high variety in customer requirements and their frequent changes while still being able to control production effectiveness and costs (JIAO and TSENG, 1999 [39]).

If traditionally the optimal technical solution to a given problem was the golden goal to be achieved by the engineering departments, so today coordination and engineering management aspects will move to the center of interest. It is only the quickest solutions, utilizing 'state-of-the-art' technology and still being technically reliable and cost-effective, which provide enough economic potential to be exploited by the company.

1.5 Knowledge and Intelligence: Natural Solutions to a Well-Known Problem

Although terms like 'The Continuous Change Management', 'Constant Adaptation to a Fast-Paced Market' or 'Costs Minimization and Execution Effectiveness' have been stipulated to be icons of the management and engineering domain, the underlying phenomena are well-known and sorely discussed in many scientific disciplines, such as Biology, Chemistry or Computer Science[3]. Dynamics, complexity and the need for precision in the business and engineering world is a natural result of a competition among multiple players for a resource that is only available or regenerated in a finite quantity: every business operation seeks to harvest its markets as quickly as possible because it faces the risk that otherwise competitors take a significant share. It will carefully utilize gained assets in order to grow and find access

[3] Chemistry, Biology and Computer Science are but a few disciplines that have discussed the effects and characteristics of evolutional developments in different experimental environments. As examples should be named Orgel's 'Test Tube Evolution', Chemistry (ORGEL, 1979 [69]), Darwins Evolution Theory, Biology (DARWIN, 1859 [19]) and Koza's genetic programming, Computer Science (KOZA, 1992 [43])

to new markets. Finally it will develop techniques to protect its markets, manipulate the existing competition and harvest more effectively – all just to gain advantage when it comes to seizing a fraction of the limited potential the markets have to offer. With the vital need for profit raised out of markets with finite potentials, and with an open population of businesses each having its own character, strengths and weaknesses that define its success or failure, any market economy fulfills the full set[4] of criteria that Charles Darwin identified as sufficient for the occurrence of natural evolution. The 'Dominating Effects', discussed and introduced earlier from an economist's perspective as phenomena in a market with monopolistic competition, are the very natural effects any evolutional development brings forth. The field of forces and the resulting dilemma every company is exposed to is the same as every living creature faces when mastering the challenge of life. By comparing the development that the economies worldwide went through to the development that brought about the variety of biological species with all their characters and skills we can only guess in which infant states today's economies are and what the future is to bring about. From the first pre-biotic molecules and the earliest living organisms, their simplicity of operation and 'boundness' to a specific environment (KUHN and WASER, 1982 [45]), to today's highly developed living creatures, showing an extraordinary dynamic behavior, an enormous precision in their actions and the ability to cope with and adapt to most complex environments, it seems that the dominance and importance of these three forces exponentially exploded throughout the evolutional development. It can be assumed certain that in the future the ability to handle dynamics, complexity and precision will be increasingly important and vital to any business operation.

Exposed to this dilemma for some 4 billion years (SCHUSTER and SIGMUND, 1982 [92]) evolution, however, has developed two highly effective concepts to address it: **knowledge and intelligence**. Paired together they provide a solution to the existing target conflict and thus facilitate effective behavior. It is their intelligence which allows highly developed systems to

[4] In general, 4 necessary and sufficient requirements that lead to evolutional developments are being discussed today:
(BANZHAF et al.,1998 [5])
1. Reproduction of individuals (=open population), 2. Variation that affects the likelihood of survival 3.Heredity in reproduction (=memory effect conserving gained evolutional experiences) 4. Finite resources causing competition.
Businesses do not reproduce in a biological sense. However, lessons learned are preserved and transferred even from one organization to another. This does result in a memory effect similar to heredity in reproduction.

adapt dynamically to their environment. Incorporated knowledge and its application enables them to develop and enact quick but still accurate responses at the 'first shot'. Last but not least it is only knowledge and intelligence that provides the necessary framework that at all facilitates the perception, analysis and management of complexity.

'Motivation' is another highly prominent and important evolutional concept. It projects and differentiates fundamental needs into well-arranged sets of high-level stimuli and orchestrates their distribution over time. Thus it is the source of a differentiated inner activation potential. 'Knowledge' and 'Intelligence' together with 'Motivation' to a big extent define a system's character. Where external stimuli and the system's own motivation are the initial triggering events, the initial driving cause for any activity, knowledge and intelligence are the mediators, providing a solution that best balances the conflicting forces and thus leading to successful behavior.

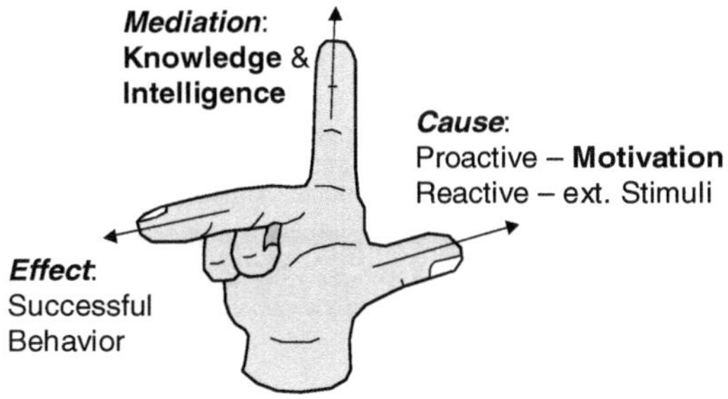

Fig. 1.6. Motivation, knowledge and intelligence – natural concepts brought forth by evolution

1.6 The Objective

Having identified 'Complexity', 'Dynamics' and 'Precision' as the three dominant and mutually conflicting effects that business as well as engineering management have to address, it is the goal of this work to provide a basic understanding and a sound theoretical framework for the given problem domain. Based on this framework feasible solutions will be proposed and important consequences discussed.

The approach taken will first derive an abstraction of the phenomena knowledge and intelligence as they are observed in the natural role models nature provides. Thereafter these abstract concepts, understandings and solutions will be transferred to the domain of business and engineering management.

After a short discussion of the state-of-the-art and the existing efforts undertaken in the area of 'Organizational Knowledge Management' in Chapter 2, the approach taken is introduced in Chapter 3. An introduction into System Theory as we apply it to the domain of business and engineering management will be given in Chapter 4. A new theoretical framework to discuss knowledge and intelligence is proposed in Chapter 5. Thereafter Chapter 6 will transfer the theoretical approach to the enterprise level, which will be followed by an illustration of technical solutions in Chapter 7. A brief conclusion will be given in Chapter 8.

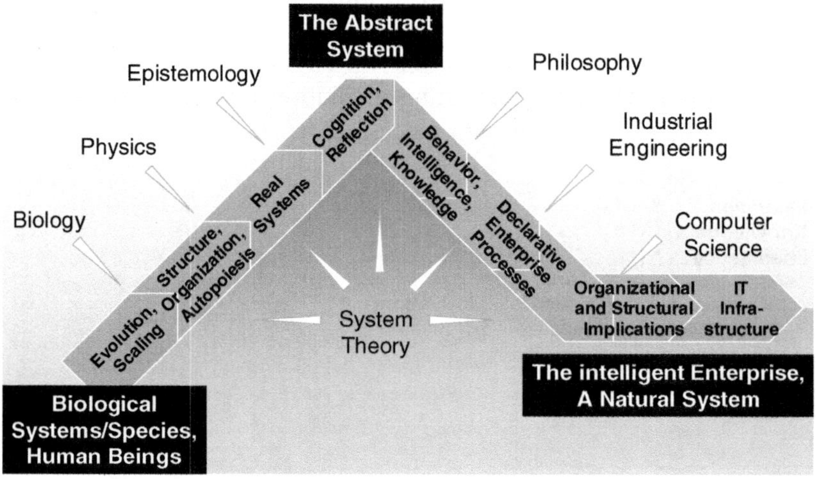

Fig. 1.7. Roadmap

Chapter 2 Background

Discussion of the State-of-the-Art

Having identified knowledge and intelligence as two key aspects to the success of enterprises competing in a market of monopolistic competition, the scientific background of managerial disciplines, which by tradition focus on these aspects – such as 'Knowledge Management' (KM) or its predecessor 'Organizational Learning' (OL) – provide valuable insights that should not escape our notice.

During the last two decades of the ending 20th century there was one managerial concept that gained increasing attention and an enormous popularity: Knowledge Management (KM). Indeed, in a time where executives faced almost uncountable new management methodologies that consultancies and research institutes day by day developed and promoted only few found a similar recognition and even fewer survived an equally long period of time. After more than 15 years of research, consulting and implementation of activities, Knowledge Management has grown mature and today seems more important than ever.

Not surprisingly there are good reasons for the sustaining interests in this management methodology. Knowledge Management provides a highly integrative approach to tackle today's managerial challenges. Historically its development was driven by efforts out of three very important management domains:

- The management and harvesting of what was identified to be the most important production factor: Intellectual Capital (IC);
- Efforts to advance the Organizational Learning (OL) paradigm and
- Efforts to extend Information Management (IM) and better utilize Information Technologies (IT).

2.1 The Intellectual Capital Discussion

Historically, the identification of a fourth production factor next to land, labor and capital can probably be regarded as the oldest root of today's Knowledge Management. According to Stewart the term "Intellectual Capital" goes back at least to 1958 (STEWART, 2001 [100]). The management theorist Peter Drucker early identified the knowledge worker as a company's 'resource' (DRUCKER[5], 1985 [26]). A scientific introduction of knowledge as a 'Production Factor' was given by the production theorist Wittmann (WITTMANN, 1979 [121]), and as early as 1987 the first conference scientifically discussing knowledge assets was held at Purdue University: "Managing the Knowledge Assets into the 21st Century", sponsored by Digital Equipment Corp. and the Technology Transfers Society. Despite its early recognition it took the discussion on Intellectual Capital until 1991 to gain momentum. In this year Stewart wrote the first of several articles in Fortune (STEWART, 1991a [98], 1991b [102], 1994 [103], 1995 [99], 1997 [101]), leading to a boost in new managerial literature published in the years to follow (BONTIS, 1996 [12]; STEWART, 1997 [101]; ROOS et al. 1998 [85]; BONTIS et al, 1999 [11]). Intellectual Capital now was discussed to be the most important production factor slowly replacing land, labor and capital (DRUCKER, 1993 [24]). Intellectual Capital, as the only source of innovation, and knowledge is becoming a more and more decisive factor in the value adding process (STEHR, 1994 [96]). The big discrepancy between knowledge-intensive companies' market values and their physical assets suggests the intangibles do play an important role, which is not captured on conventional balance sheets (WAH, 2000 [114]). In the mid-1990s numerous approaches were proposed to capture, evaluate and even balance intellectual capital, leading to the first ever published intangible asset figures by the Swedish company Celemi in 1995. Among the most prominent methodologies to evaluate intangible assets are EVA® (Economic Value Added[6]), BSC (Balanced Scorecards) and HRA (Human Resource Accounting) (BONTIS et al. 1999 [11]). Measuring intangible assets was one problem, but how to manage them was the big question driving research in the late 1990s. In order to educate and achieve the necessary mind shift Sveiby introduced a business simulation with an explicit focus on intangible assets called Tango that should help executives and managers to gain the necessary sensibility towards Intellectual Capital in their day-to-day work (SVEIBY, 1997 [105], SVEIBY and MELLANDER 1994 [104]).

2.2 Information Management and IT Driven Developments

Ever since computers found their way into business organizations they were used to collect and evaluate data available along business processes. Processed and edited by the controlling departments they were turned into valuable

[5] Which was originally published already back in 1973.
[6] EVA® is a registered trademark of Stern Stewart & CO.

information (WITTMANN, 1959 [120]), which is the basis for all management decisions. For many years making sure the right information is available at the right time has been the main duty of controlling and information management (BAUMOEL, 1998 [7]).

However, the picture changed with the rapid development in IT infrastructures. As computers grew faster and the available network technologies improved, their application changed as well: in addition to process monitoring, they now serve communication purposes, documentation and archiving, process management and even computer assisted project management. Just as 'Data' was lifted to the higher 'Information'-level by processing and editing, IT experts hope to achieve yet another metamorphosis by embedding information into a larger context, structuring and classifying it, in order to make it easy to digest for employees who need to build up personalized 'Knowledge' (ROBERTS-WITT, 2000 [83]). Today's IT tools are not just programmed to capture data and process it to information, but they even support the management of documents, ideas and creativity. Capturing not only process data but content created by people and their interactions, today's Knowledge Management tools provide clearly more than was ever expected from information management (GLADSTONE and KAWALEK, 2000 [31]). In many cases information technologies indeed did play a major role when it came to implement Knowledge Management. The need for well structured and organized electronic archives and databases that provide essential information for the employees to carry out their day-to-day work seems to exist throughout almost any business operation. For this reason IT professionals often define these electronic repositories as the companies' "corporate memory" (KINNI, 2000 [41]).

2.3 The Pragmatic Extension of Organizational Learning

The third pillar completing the fundament of Knowledge Management grew out of the theoretic framework once developed for the 'Learning Organization'. Probst and Romhardt propose to understand Knowledge Management "as the pragmatic extension of the ideas of 'Organizational Learning'" (PROBST and ROMHARDT, 1997 [78], p. 1). To an increasing degree, KM experts realize that the cultural shift towards an interactive learning environment is essential for any Knowledge Management implementation (WAH, 2000 [114]). Efforts to form an open culture that honors success and learns from failures, ongoing educational programs and professional training as well as a strong dedication to continuous improvement are integral parts of almost all holistic approaches to Knowledge Management.

The Organizational Learning domain provides a sound foundation of scientific publications, in which many of the KM approaches are rooted. Nonoka and Takeuchi identify four prominent knowledge conversions to be at the center of the organizational knowledge creation process: combination, externalization, internalization and socialization (NONOKA and TAKEUCHI, 1995 [66]). Learning within an organization is understood as a process established by interactions between the individuals and the social group they belong to. In these interactions individuals express and formulate their knowledge, thus generating explicit[7] knowledge; knowledge of multiple experts is combined; newly generated insights, methodologies and practices are incorporated as behavioral schemes (implicit knowledge); and behavioral schemes are imitated and adapted by other group members.

In addition to the traditional organizational learning approaches Knowledge Management provides a strong focus on substance and content (PROBST and ROMHARDT, 1997 [78]), giving it a clear target definition next to a strong commitment to personal education and professional training.

2.4 KM Models – Approaches to a Holistic Knowledge Management

The beauty of the integrative character, with three powerful managerial concepts being integral parts of the holistic Knowledge Management approach, also inherits one of its biggest difficulties: finding the right strategy to select and implement the right set of activities. Driven by pragmatic motives the research community has proposed several approaches most of which are built on so-called Knowledge Management Models (BULLINGER, WOERNER and PRIETO 1997 [14], PROBST, RAUB and ROMHARDT 1999 [77], GREENWOOD 1998 [33], HEISIG 1998 [38], DAVENPORT and SMITH, 2000 [20]). These models most commonly organize activities and measures to be implemented in a two-dimensional matrix. The first of its

[7] The separation between explicit and implicit knowledge is highly prominent throughout the KM community and in particular stands at the center of Nonoka and Takeuchi's knowledge generation process. Here explicit knowledge is largely understood as facts, methods, principles, techniques, etc. that can be articulated or even have been articulated. Implicit knowledge on the other hand is knowledge that cannot be articulated, such as behavioral patterns, internalized procedures, cultural habits, etc. However, this distinction is subject to heavy discussions. So, on the one hand most of the cognitive scientists would regard anything outside the human brain (e.g. formulated procedures, facts, formulas, ...) as information but not as knowledge. On the other hand, non-articulated knowledge is sometimes even further differentiated into tacit knowledge, which cannot be articulated at all, and implicit knowledge, which in principle can be articulated (NICKOLS, 2000 [64]).

dimensions is an activity classification based on the activities' character with regard to the functional role they play in the Knowledge Management effort. This classification in so-called Knowledge Management 'Building Blocks' (PROBST and ROMHARDT, 1997 [78]) includes classes such as 'Knowledge Acquisition', 'Knowledge Creation', 'Knowledge Distribution', 'Knowledge Application' and many more. The second domain classifies affected organizational domains in which Knowledge Management efforts have to be undertaken. They are often called the 'Design Dimensions' (BULLINGER, WOERNER and PRIETO 1997 [14]) or 'Design Fields' (HEISIG 1998 [38]) such as 'Human Resource Management', 'Corporate Culture', 'Information Technologies', 'Leadership', 'Controlling' and others. The granularity implemented on each dimension varies greatly from model to model. Some use up to eight 'Building Blocks' in only three 'Design Fields' whereas others use only four 'Building Blocks' but define six 'Design Fields'. The KM building blocks are usually arranged in one or even more cycles to represent closed feedback loops – starting with the definition of goals, ending with the validation of achievements and providing the necessary control parameters to fulfill managerial requirements.

These Knowledge Management models are a pragmatic conceptualization of activities – they decompose the management process in logical phases and provide a search grid for companies to analyze deficiencies and start new efforts (BULLINGER, WOERNER and PRIETO 1997 [14]). According to Hansen, Nohria and Tierney each of the implemented activities either follows the 'Codification Strategy' – gathering, structuring and contextualizing information in knowledge repositories and distributing it from there reactively – or the 'Personalization Strategy' – involving training and educational programs to distribute and build up personnel knowledge proactively (HANSEN, NOHRIA and TIERNEY, 2000 [34]). In an effort to better target implemented activities the KM community has learned to further classify knowledge itself. So Schueppel proposes eight fundamental knowledge types (SCHUEPPEL, 1996 [91]), whereas Mertins, Schallock and Arlt break down their classification into even 22 distinct types (MERTINS, SCHALLOCK and ARLT, 1994 [61]). Even more common is the classification of knowledge into mutually exclusive property pairs such as implicit vs. explicit knowledge, formal vs. informal knowledge, individual vs. collective knowledge and many more (ROMHARDT, 1996 [84]).

2.5 The Traditional Knowledge Management

In one of his recent publications Peter Drucker argues that the 50-fold rise in productivity of the manual worker was one of the biggest achievements the 20th century brought forth. Drawing an analogy, he concludes that finding the right managerial concepts to bring about a similar rise in productivity of knowledge workers is one of the most important challenges the 21st century has to master (DRUCKER, 1999 [25]). Exactly this is the goal Knowledge Management strives to achieve. Despite all the brilliant ideas proposed the question how this is to be achieved is still subject to heavy discussions. Where Wha identifies

- "Capturing, storing, retrieving and distributing tangible knowledge assets, such as copyrights, patents and licenses."
- "Gathering, organizing and disseminating intangible knowledge, such as professional know-how and expertise, individual insights and experience, creative solutions and the like."
- "Creating an interactive learning environment where people readily transfer and share what they know, internalize it and apply it to create new knowledge."

(WHA, 2000 [113], p. 308) as the essence of Knowledge Management, Johnson[8] simply proclaims that "knowledge management is all about capturing and using know-how" (JOHNSON, 2000 [40], p. 85). In an article published in 1999, Malhotra has given no less than ten different definitions for Knowledge Management from leading KM authorities (MALHOTRA, 1999 [55]). With the enormous popularity and an estimated five billion U.S. dollars spent per year to KM consultants alone in the United States (SRIKANTAIAH, 1999 [95], p. 10) soon hit the downside of the glamorous rise: the increasing competition within the Knowledge Management community led to disputes, tearing apart what once was an integrative approach and severely damaging its beauty. Knowledge Management IT experts sought for a clear distinction between Knowledge Management and Information Management tools, claiming that only they offer the right tools to manage knowledge (GLADSTONE and KAWALEK, 2000 [31]). Training experts criticize the IT

[8] The sentence "Knowledge management is all about capturing and using know-how, usually with an internal focus." (JOHNSON, 2000 [40], p. 85) appears right in the abstract section of Johnson's article in the 'Knowledge Management Yearbook 2000-2001', where it is unclear whether it was written by Johnson or the editors J. W. Cortada and J. A. Woods. Exemplifying different notions of understandings and approaches, this should not hinder the discussion.

domain for merely providing mechanistic solutions without great value (GORDON, 2000 [32]).

An even bigger problem of traditional Knowledge Management is the absence of a sound scientific foundation. Knowledge Management was born out of activism, empirically classifying and structuring the problem domain and conceptualizing activities. However, knowledge, the substance and focus to which all efforts are aligned, is barely understood! So Nickols still described knowledge as “a very slippery concept” (NICKOLS, 2000 [64], p.20) in the Knowledge Management Yearbook 2000-2001, making it difficult – if not impossible – to discuss Knowledge Management scientifically. The established KM Models, which are often adopted as a theoretical framework, inherit the same deficiencies. Probst and Romhardt, both pioneering the development of KM Models, clearly point out that their Building Blocks are merely a conceptualization of activities following no other external logic and emphasize that the right model for Knowledge Management does not exist (PROBST and ROMHARDT, 1997 [78]).

The traditional approaches borrow parts of their conceptual framework surrounding the management of knowledge on an organizational level from Epistemology, ignoring the fact that this philosophical discipline by definition limits its discussion to human knowledge and human understanding (KUNZMANN, BURKARD and WIEDMANN, 1991 [46], p. 119). Moreover, the modern epistemological understanding of the self-referenced development of human knowledge (PIAGET, 1983 [70]) clearly calls constructs such as Nonoka and Takeuchi’s organizational knowledge creation process into question (EBERL, 2001 [28]).

With the growing awareness that

- Knowledge Management programs tend to be costly while still not necessarily return investments quickly,
- successful implementations in one enterprises are not necessarily transferable to a different organizational environment and
- the general predictability of results returned by KM efforts is low

the desperate need for a better scientific understanding is evident.

2.6 New Scientific Approaches

With Knowledge Management getting beyond the point where it could be seen as a management fashion like many others that come and go, it is slowly growing mature. One strong indicator is the fact that indeed more and more scientific discussions are published on this domain.

Recently Ocasio introduced an adaptation of the 'SERSTS-Model' (the situation-enactment-retrieval-selection-transmission-storage model of hu-man information processing introduced by Weick (WEICK, 1979 [115]) to describe how organizations think. In his publication he describes how the organization structures and regulates individual cognition by proposing to understand the organization as a network of situations in which individuals think (OCASIO, 2001 [67]). Sitkin in accordance sees cognition on an organizational level as "a rubric under which is included collective, aggregate[d] patterns of individual cognition within organizations" (SITKIN, 2001 [94], p. 76).

As early as in 1995 Krogh and Roos introduced a new perspective to viewing and describing organizations (KROGH and ROOS, 1995 [44]). Their approach is based on a system theoretic view on organizations. They emphasize the central role of language and languaging. Knowledge emerges while individuals play language games that follow established rules. The concept of scaling was introduced to conceptualize the similarity of knowledge build up in organizations and individuals. Again it is the individual that takes a central role, playing the language game, taking decisions and balancing what Sitkin probably would call 'Exploration' and 'Mastery'[9]. Just as in the traditional KM approaches the demand arises for a solid organizational structure that facilitates directed learning.

All cognitive attempts to address KM issues by describing organizational cognition and thinking as an aggregation of human cognition or as a network of situations in which individuals think seem to be a difficult endeavor, which – up to now – leave the KM community with few practical insights. A similar situation faces Krogh and Roos' approach (BERTELS, 1997 [9]). In particular, some of its central concepts such as the 'Language Games' do not seem feasible in the engineering or manufacturing domain.

2.7 Conclusion and Consequences

On the one hand, the 'pragmatic-activistic' approaches to Knowledge Management lack a scientifically sound theoretical framework. Most of the theoretical approaches on the other hand simply fail to address the problem domain on the implementation level.

---

[9] The balance between exploration and mastery is discussed as a central concept of cognition (SITKIN, 2001 [94]). Levinthal and March use exactly this balance to conceptualize intelligence (LEVINTHAL and MARCH, 1991 [49], MARCH, 2001[57]). In a similar fashion Krogh and Roos identify the balance between the creation of new knowledge and the realization of achievements as critical for the company's success.

If one analyzes traditional Knowledge Management efforts it is conspicuous that the individual occupies the focus of interest. Knowledge Management tries to condition the individual employee to sharpen his skills, decision making and his personal work effectiveness in educational activities as well as by supplying the right data and information, structured, contextualized and ready to be incorporated right at the moment of need.

Although this may seem obvious for most pragmatic KM efforts, it is highly surprising that most theoretical approaches follow the same path by introducing organizational cognition, thinking and knowledge as phenomena created by and emerging from individual cognition, thinking and human knowledge. The organizational forms are not treated as distinct phenomena but are assumed to be aggregations of their human counterparts.

This is in particular surprising as the idea of a collective mind has been discussed and dismissed by Allport as early as 1922 (ALLPORT, 1922 [1]) and later by Douglas in 1986 (DOUGLAS, 1986 [23]). Even earlier, Taylor already destroyed the myth of personal skills by even dismissing their existence (TAYLOR, 1911 [106], DRUCKER, 1999 [25]). What the skilled craftsman carries out is a series of actions, motions and decisions that Taylor separated and reorganized in tasks. It was the management in tasks and the task coordination which brought about the enormous rise in productivity. Laying the roots for modern Industrial Engineering, his approach proved to be valid throughout all industries and gave rise to the emergence of industrial and developed nations (DRUCKER, 1999 [25]).

Surely time has changed. Today's efforts aim at increasing the productivity of the knowledge worker, not the manual worker, making it difficult to adopt Taylor's theories in the strict sense, but aren't his underlying ideas still valid today? At his time, Taylor's ideas were highly radical and unconventional, but even today it should not be ruled out that again a new radical line of attack is required to push ahead!

In this work a new approach is proposed, looking at the enterprise as a natural and real[10] system and at how this system as a whole sustains and manages the forces to which it is exposed. Given the picture painted in chapter one, the question on hand is: What phenomenological consequences does the evolutional development of enterprises in their markets have? Or in other

[10] A 'real system' in the sense of sociological System Theory is a system that defines its own border, leaving it not up to the observer what to regard as part of the system and what not. The system as a real entity existing without prior construction by the observer is the central idea in Luhmann's System Theory (REESE-SCHAEFER, 1992 [81]).

words: What phenomena characterize a system that successfully balances dynamics, complexity and precision? Similar to Krogh and Roos this is a system-theoretic approach. However, the system-theoretic framework was developed and adopted with an emphasis to cater for the engineering and engineering management domain.

Observing the system ‘Enterprise’ as a natural system exposed to and formed by evolutional forces inevitably demands to understand ‘knowledge’ just as a part of a greater whole: the Intelligent System. Intelligent behavior, or the ability to successfully adapt to environmental changes and exploit unbalances, last but not least includes more than only knowledge. It also requires a system body providing all necessary degrees of freedom and their associated actuators that facilitate adaptation and allow the enactment of highly diverse activities. Discussing knowledge isolated from an intelligent system would be a rather meaningless endeavor. Asking the question of what emerged first, the adaptive system and its framework of activities or knowledge that guides these activities, is like asking the question: What was first, the hen or the egg?

Chapter 3 The Approach

A Discussion of the Proposed Approach, Its Scientific Grounding and Direct Consequences

The proposed approach treats the enterprise as a natural living system. In management sciences, approaches that describe and analyze enterprises as a living organism are widely known as so-called 'Systems Theoretic' approaches (SILBIGER, 1999 [93]). All of them find their roots in 'General Systems Theory', which was proposed by the biologist Ludwig von Bertalanffy in the 1940s (BERTALANFFY, 1968 [8]). Typical applications in the management sciences use systems theory to carry out organizational analysis by comparing the organizational bodies of an enterprise to the subsystems of a paramecium. Systems theory conceptualizes high-level methodologies to understand and describe complex systems in their unity. These ideas later brought forth several scientific disciplines, among which 'Cybernetics' (ASHBY, 1960 [4]) and later sociological 'System Theory' (LUHMANN, 1984 [53]) are the most prominent.

3.1 The Biological Comparison

> *Whenever biologists try to formulate definitions of life, they are troubled by the following: a virus; a growing crystal; Penrose's tiles; a mule; a dead body of something that was indisputably alive; an extraterrestrial creature whose biochemistry is not based on carbon; an intelligent computer or robot. – William Poundstone, The Recursive Universe (1984); (MORALES, 2001 [63])*

Discussions about the definition of life and living system have a long and prominent history, involving scientists from all disciplines and nations. Introducing an assumption which eventually gets quite close to the gray zone dividing the 'living' from the 'nonliving', it is indicated to start the discussion staying clear from the borders. For this purpose the paramecium again should serve as a starting point. As systems consisting of only one single cell, the paramecia are probably the simplest systems which are still agreed upon as being natural living systems. Analytically, they can be seen as an aggregation of biomolecules that actively process nutrition to grow and reproduce.

Exposed to a harsh and highly competitive environment evolution favors those organisms that best adapt to their environment, most efficiently utilize their resources and most successfully reproduce.

Reproduction obviously occurs right where the reproducing system is located and only at a point of time at which the reproducing system itself is still alive. Consequently, a clustering of living systems over the spatial and temporal dimensions occurs, naturally leading to the formation of communities. Once again exposed to evolution for generations, these communities change their appearance, improving their internal integrity and overall effectiveness as forced by selection and extinction. As a result, higher organisms that have strongly integrated their constituents – body cells – form. Each organism as a whole gains a better access to resources, processes them more efficiently and is less dependent on specific environmental conditions than any of the single cells would be as sole players. Through the integration of elements to a larger system and coordination of their activities the organism has access to new degrees of freedom, a higher overall effectiveness and adaptability. Having even arranged production mechanisms on the organism's level – reproduction of the organism as a whole – relieves it from the need of a random clustering to generate a new entity and allows it to preserve the structure and organization that have been proven as best practices. This way higher organisms, plants and animals, came into existence and developed – last but not least as they provide a clear evolutional advantage over the aggregating organisms, the cells.

With the continuous reproduction of higher organisms a similar clustering inevitably occurs again. This time it is colonies of higher organisms, social groups and states that emerge and develop. Regarding ant states and bee colonies as unitary living systems – or as organisms on an even higher level – is not uncommon. However, it usually requires some persuasion if social systems such as enterprises, societies, organizations and the like are introduced as living systems. Without doubt there are differences between the biological systems that are commonly accepted as living and social systems like enterprises. The two most important differences are:

- The system's constituents are not necessarily reproduced within the system itself – employees e.g. are not 'reproduced' within the enterprise.
- Social systems usually do not reproduce as a unitary whole – the enterprise e.g. does usually not reproduce its kind[11].

[11] Even though company spin offs (s. Chap. 3.3 b)) are getting increasingly popular and they clearly qualify as a reproduction that preserves structure and organization (reproduction as a unitary whole), they still are more the exception than the rule. Up to now, most companies form spontaneously as typical start-ups.

On the other side does the system enterprise very well fulfill most of the criteria that commonly identify living systems:

Enterprises, e.g., …

- show a responsive behavior to stimuli from the environment or from inside themselves, leading to a behavior that promotes their own continuation.
- are inherently active. It is their ongoing set of internal processes that shows and defines their existence.
- have a characteristic order that is actively preserved and thus resists the inevitable entropic decay.
- are open systems, exploiting unbalances by exchanging material and energy while at the same time increasing the entropy of their environment.

All this suggests that, despite the two arguments[12] mentioned earlier, the enterprise should be regarded as a natural living system.

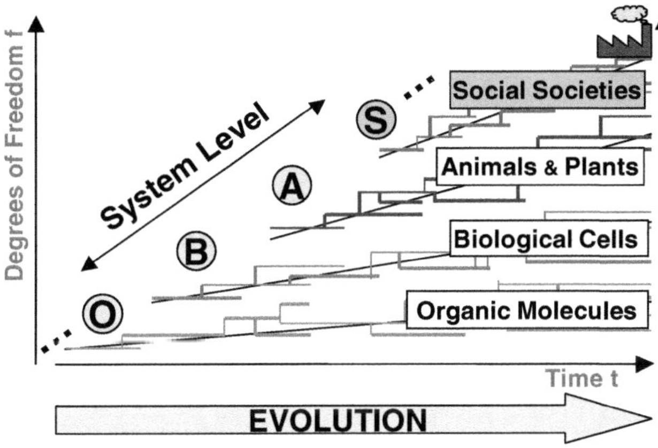

Fig. 3.1. System levels

[12] Both arguments are easily resolvable. The first one, arguing that the social system's constituents (e.g. the employees) are not reproduced within the system, loses its weight as humans/employees are not to be regarded as the system's structural elements (s. Chapter 4.5). The second one, arguing that the system does not reproduce as a whole, is resolved in Chapter 3, 3.3-b. Furthermore Morales shows that both arguments cannot be regarded as strong definitional sentences (MORALES, 2001 [63]). An easy example against the second argument is the mule: It is definitely a living system although it can't have offspring itself and thus can't reproduce as a whole. This and more examples can be found in MORALES, 2001 [63].

Having introduced the evolutional path that implies the development of organisms, aggregating constituents over different evolutional stages and complexities, there is almost a demand for a classification that allows two distinct systems exactly based on this criteria: the 'System Levels'. We identified the 'System Level' of a system by the type of constituents it aggregates. Given the four system levels shown in Fig. 3.1, 'Organic Molecules' (0), 'Biological Cells' (B), 'Animals & Plants' (A) and 'Social Societies' (S), body cells e.g. are obviously to be classified under B, human beings are level A systems and enterprises are systems on System Level S, 'Social Societies'. The higher the System Level of a system, the larger the number of systems on lower System Levels that have been nested in order to integrate it.

Note that this classification represents a pragmatic conceptualization rather than a scientific taxonomy, introduced merely to reduce the communication efforts when discussing effects and concepts linked to this hierarchical composition that evolution brought forth. Note furthermore that the given line of system levels – 'O', 'B', 'A' and 'S' – can and eventually needs to be expanded on both sides. Doing so will lead the line to expand into areas that can clearly be identified as 'nonliving' without breaking the logic of classification.

> *Nature proceeds little by little from things lifeless to animal life in such a way that it is impossible to determine the exact line of demarcation ... – Aristotle, The History of Animals, (350 B.C.) book viii: part 1. (ARISTOTLE, 350B.C. [3])*

Following the notions introduced above a system theoretic approach is proposed that assumes **enterprises to be natural living systems** on the System Level **Social Societies ('S')**.

Without stepping deeply into systems theory the approach itself already implies a number of consequences that should be discussed briefly in the remains of chapter 3.

3.2 The Scales of Learning

Regarding evolution as an ongoing 'Continuous Improvement Process' two major improvement-levers can be identified to be at work simultaneously:

1. The continuous enhancement of systems within all system levels by gearing up their internal integrity, efficiency, adaptability, etc.
2. The step-by-step increasing of System Levels, each providing numerous new degrees of freedom.

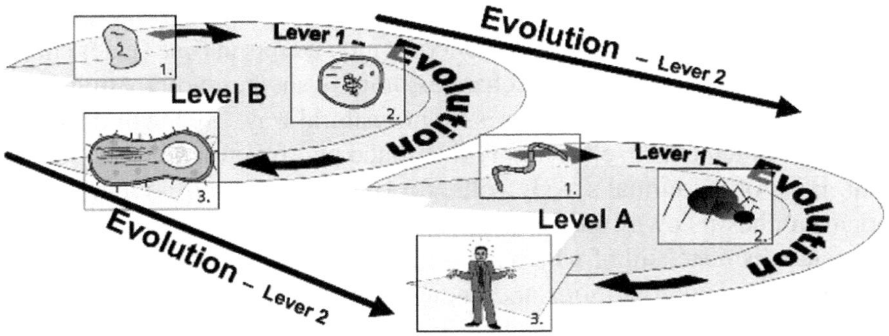

Fig. 3.2. Evolution and improvement-levers

Not surprisingly the resulting effect is identical to the 'Jumping S-Curves of Innovation' discussed in chapter 1. Beginning virtually from zero, right after the jump to a new level, the evolutional mechanisms enhancing the systems' integrity, effectiveness, adaptability, etc. have to start all over again. With the disadvantage of a significantly shorter development time one would expect high-level systems, forming much later than low-level systems, to lag behind in their overall evolutional development. This is indeed the case for very early stages. However, the evolutional development of high-level systems progresses much faster than for low-level systems and thus quickly overcompensates the initial disadvantage.

Where improvements in simple and low-level systems require selection and extinction – learning thus happens slowly over generations (**evolutional learning**) – highly developed systems on higher levels are able to adapt their behavior schemes and learn within their own lifetime – learning by refining soft-wired behavior patterns based on their success (**intelligent learning**). It is remarkable that this adaptability and learning in higher-level systems happens although their constituents themselves are unable to learn intelligently. The higher-level systems' adaptability, ability of learning and intelligence not only by chance exceed those of its lower-level constituents by scales, but this is rather an evolutional necessity: it is the only possibility to gain a competitive advantage against lower-level systems acting as sole players and thus it is the only reason for the existence of high-level systems.

Consequently we have to expect intelligent enterprises to adapt and learn faster than any of its employees would be capable of. The employees cannot carry the learning and adaptation process of the organization – although they do enact it.

Inevitably this raises a question of interfacing between system enterprise and employee: if the company's behavior constantly adapts in time scales shorter than the employee can realize changes, how is she/he participating in the enactment? Biology resolves this issue in a natural way, as each of the organism's constituents is itself a highly integrated and functionally autonomous unit. Each of an animal's body cells, for example, is itself an autonomous organism, has its own functional processes, its own subsystems to manage its metabolism etc., all of which are – given the case that it is provided with the right physical conditions and enough nutrition to run its internal processes – functionally independent from its environment (the animal body). Nevertheless, given the right stimuli the cell will enact its part of the global process. The stimulus is the trigger upon which the cell responses, by enacting whatever is necessary. The cell is never told how to enact its parts, only what to enact! It is the stimulus that is providing the cell with a notion about 'what to achieve' – it provides the **'logic'** to 'what' it will enact – whereas the how or the **'control'** of the enactment is left completely up to the cell. Implemented consequently, a global stimulus to an intelligent system representing the 'what to achieve' will trigger the auto-assembly[13] of a response process including numerous body cells that enact the system's overall behavior. It is an auto-assembly, as the stimuli traverse from cell to cell, solely based on how the individual cell handles a given stimulus. Given a slightly different stimuli constellation, a highly different response behavior may be enacted. Given further infinitesimal changes in the response behavior of individual cells (e.g. brain cells that slowly grow new synapses as a learning effect), again a highly different process may be enacted even when experiencing the same stimuli constellation. This intelligent processing, or, as it is introduced later, **'Declarative Processing[14]'**, allows the higher-level organism to achieve adaptation and learning within shortest time scales while its constituents only insignificantly refine their behavior.

---

[13] Given a certain stimulus constellation, a response process and with it the response behavior of a higher-level biological system is auto-assembled in the sense that the stimulus to any given neuron is exclusively processed and handed on to other neurons based on its own internal functional rules. The stimuli are thus processed without influence or control exerted by the system itself but only by its constituents. The stimuli happen to address corresponding motor regions to enact behavior based purely on how they were processed from neuron to neuron. This does not mean that behavior is hardwired or predefined as neurons may grow new synaptic connections or break up old ones (soft-wiring). Almost infinitesimal changes in the way they hand on the stimuli from one to another may result in a highly different response process.

[14] Declarative Processing is introduced in greater detail in Chapter 5.5. For more details please see the glossary entry and read Chapter 6.4.

The key to this effect is the functional independency or self-referentiality of the constituents. It is only this functional closeness and self-determination of the constituents – bringing in their pre-constituted complexity – that facilitates the complexity of the organism they live in.

The difference between addressing an employee with 'What to achieve' rather than with 'How to go for one's work' is exactly what Drucker identifies as the characteristic difference between 'Knowledge Worker' and 'Manual Worker' (DRUCKER, 1999 [25]). With every employee objectively fulfilling the requirements – being her-/himself a self-referenced, self-determined individual (PIAGET, 1983 [70]) – it is only a question of procedural integration in the network of company processes that determines whether the employee is a 'Knowledge Worker' or not. Where the employees clearly meet all necessities, most – if not all – enterprise processes fail to effectively integrate the 'Knowledge-Workforce', which is today experienced as a striking deficiency in the knowledge worker's productivity (DRUCKER, 1999 [25]). Without doubt solutions are not a question of improved task coordination, communication tools, TQM, Kaizen or Business Process Reengineering. Rather, a radically new definition and understanding of processes is required here: an understanding that facilitates the realization of 'Declarative Processing' and the effective integration of employees as 'Knowledge Workers'.

3.3 Systems Exposed to Evolution

As all other living systems, enterprises are exposed to evolutional developments and the continuous improvement through natural selection and extinction. Thus every enterprise faces a very natural physical and biological reality implying a number of significant consequences.

As a physical system that persists by actively resisting the entropic decay it is forced to consume energy as well as building blocks, both of which are processed and also constantly released to achieve its fragile stability. The existence of unbalances in its environment and their exploitation are vital needs. From a biological point of view a number of other aspects, like the need for growth and reproduction or staying healthy, indicate further requirements to persist evolution. Highly developed systems have survived others for their improved abilities to

- sense unbalances, opportunities, perturbations and threats.
- quickly react and adapt to changes in their environment.
- anticipate or even predict developments.

- actively influence and purposefully form their environment, provoking opportunities and conditioning it in a favorable way.

All those abilities require a number of physical properties or abilities that systems on all discussed System Levels slowly developed throughout the evolutional process. These include:

- Processes to integrate new building blocks and achieve structured growth.
- Processes that arrange the sharing and distribution of resources.
- Communication means among the players that constitute the system.
- Reproduction that conserves structures, organization and processes.
- The development of procedures that realize effective task sharing.
- Structures that realize declarative processing to reach a significant level of complexity, guaranteeing an optimal adaptability.

This short list can be easily extended or further broken down, but already in this simple form it raises some significant implications:

a) Throughout the evolutional development systems emerged showing a higher and higher intelligence, which distinguished themselves from simpler and less intelligent systems through the simple fact that they had a number of physical properties and abilities improved. A system's intelligence or capability of learning and adaptation are defined by its physical setup! Even the brightest fox by far will not be nearly as intelligent as one would expect a human being to be – just because the physical framework and organizational setup of its body doesn't support it.

The intelligent enterprise primarily depends on sound physical conditions – the structural and organizational setup – rather than on virtues, soft skills and corporate culture. The latter are of high importance and great value to any managerial efforts and thus are in general also important for Knowledge Management. However, it is difficult to comprehend that soft factors should play any extraordinary role in KM efforts. Their stipulation as icons in many KM efforts seems to provide an escape hatch for consultants rather than a shift towards a more intelligent enterprise.

b) Today's enterprises have only poorly developed those significant physical properties and abilities mentioned above. So, most businesses do not start up through a controlled reproduction but rather through the process of random clustering. Only recently have more controlled segregation and reproduction processes – from profit center to divisions and later spin-offs – that conserve structure, organization and processes, become more common. The reason why enterprises do not reproduce is not that this is phenomenologically impossible, but they are just not developed enough. Another point is the strong 'Taylorian' approach most manufacturing enterprises still follow, leaving little room for intelligent or declarative processing.

It indeed seems that enterprises as 'Level S' systems still lag behind in their overall evolutional development, having the disadvantage of a significantly shorter development time compared to other systems on lower system-levels. Today's enterprises still are at the very beginning of their evolutional development, which shall give rise to great expectations for the years to come.

3.4 Scaling and Its Consequences

Krogh and Roos introduced the concept of 'Scaling' to discuss and describe the similarity and coupling of knowledge generation in individuals and organizations (KROGH and ROOS, 1995 [44]; BERTELS, 1997 [9]). The observation of mechanisms and effects in individuals should help to understand the requirements on the level of an organization. What the employee is to the organization is the body cell to the human, what language is to the organization are hormones, neurotransmitters, etc. to the human body and so on. 'Scaling' gains its scientific validity by the fact that the evolutional mechanisms are identical on each System Level, leading to matching functional, structural and organizational consequences. It is exactly these physical consequences (and no others) that can be compared by applying the concept of 'Scaling'.

Human knowledge, human understanding and learning have been described by Epistemology, Education and Psychology. Their physiological effects, however, are the domain of Neurobiology. The way a human being reacts to stimuli that are perceived by a number of sensory neurons is defined by the way these stimuli are processed declaratively from neuron to neuron, over possibly thousands of synaptic connections, until they reach the muscle cells that finally enact the response behavior. All learning – no matter how complicated its psychological description may be – is finally the result of a finite set of neurons that have grown synaptic junctions, broken contacts or in any other form changed the way they process stimuli from one to another. When a human being learns, she/he does neither condition her/his body cells (like the speed of signal transfer, its failure rates or the critical threshold a neuron needs to fire) nor does she/he improve any of the neurotransmitters or hormones involved. What changes is only the interplay of neurons and their coordination of activities.

Applied to the organization this proposes that an intelligent enterprise does neither learn by conditioning its employees, improving their personal knowledge, skills or decision-making (as all traditional KM efforts strive for), nor does it learn by improving the language or its perception (as implied by Krogh and Roos' language games). The intelligent enterprise learns by redefining the coordination of activities that are enacted by the employees.

Any of the traditional KM and OL efforts that try to condition the employees compare to learning efforts of a human who tries to condition her/is body cells in order to learn. This may be done e.g. by students who take a rest before an exam to be refreshed and alert, but continuous resting would not lead to any learning!

A second important consequence is imposed by the fact that consciousness and cognition in the human body are not rooted in consciousness of any of its body cells. Human cognition is not an aggregation of cognitive phenomena existing on the level of its cells. This strongly implies that cognition on an organizational level is not to be understood as an aggregation of individual cognition, just as reflection on an organizational level cannot be regarded as a network of situations in which individuals think. This strongly opposes most of the proposed cognitive approaches (s. Chap. 2.6).

3.5 Phenomenological Consequences

Understanding a company as a natural living system opens new possibilities for the phenomenological discussion of 'organizational knowledge' and 'organizational intelligence'. Both now can be discussed as phenomena of a natural living system in an effort that first tries to describe both, knowledge and intelligence, on an abstract level before transferring them to the system enterprise.

Motivated in particular by Epistemology, which exclusively discusses human knowledge and human understanding, knowledge today is often understood as a phenomenon that exists exclusively in human brains. From the perspective of the natural sciences, however, there are no strong indications for this understanding. Knowledge is a concept that improves or even facilitates a system's ability to effectively adapt to its changing environment. Thus it is an imminent consequence, demanded by evolution's central principles. As evolution itself is a rather smooth process, reusing its concepts over and over and improving them in almost infinitesimal steps, there is no rationale behind an assumption that treats knowledge as a purely human phenomenon. Rather, does knowledge as well as intelligence, in one form or another, probably exist in all systems and on all System Levels, and only have evolved over time to what we experience today as human knowledge and human intelligence. However, it is remarkable that knowledge in other systems or on other System Levels has never been identified as such! Obviously it drastically changes its phenomenological appearance from system to system and System Level to System Level, which prevented this phenomenon from being identified as knowledge anywhere outside the human domain.

Consequently, one has to expect both phenomena, knowledge and intelligence, to have a highly different appearance on the organizational level. Organizational knowledge and organizational intelligence definitely will look very different from their human counterparts. Approaches that conceptualize organizational knowledge as something close to human knowledge, expert knowledge, formal knowledge, analytical knowledge and the like should be critically called into question.

Chapter 4 Theory of Real Systems

A Discussion of Important Scientific and Philosophic Concepts from the Perspective of Engineering Management

Before the phenomenological character of perception, cognition, reflection and finally knowledge on an organizational level can be understood and described, valid generalizations should be discussed on the level of abstract systems. For this purpose, system theoretic concepts as discussed by Maturana, Varela, Luhmann, Whitaker and others are introduced in this chapter. System Theory is of particular value to the discussion of organizations and enterprises as it – in its later stages – intensively discusses social systems, their functional principles, developments, behavior, response and action patterns. Even though System Theory in the past decade has mainly been advanced and developed by Sociology and Philosophy, the presented approach attempts to introduce it from the perspective of the Natural and Engineering Sciences. This does not only give new notions and a shift of focus towards the natural living system 'Enterprise', but it is also by far easier to understand for anyone who does not have a philosophic background. System Theory provides an extensive theoretical framework. In this work the effort is undertaken to limit the discussion to only those elements that are of high relevance to the later discussion of system knowledge and intelligence. Still, the question of why we do need this philosophic discourse at all remains. Indeed, multiple new and complex terminologies as well as numerous highly abstract concepts are introduced and complicate the initial discussion. However, philosophy plays an important role as soon as results and findings from multiple disciplines have to be merged to form a greater whole. Philosophy is the common root to all scientific disciplines. It provides a sound definitional framework that can be shared for a wide set of terminologies and, last but not least, it also provides numerous highly abstract and fundamental theoretical frameworks for issues that are discussed under different aspects, perspectives and assumptions by many scientific disciplines. In the analysis of enterprises, their behavior, intelligence, knowledge, reflection and evolutional development involving valuable insights from multiple disciplines, such as Management Science, Industrial Engineering, Sociology, Biology, Physics, Epistemology, Cybernetics and many more, Philosophy plays an essential role.

This chapter aims at introducing the following three highly important concepts that are essential for the later discussion:

- A new understanding of **processes**[15] that is conceptually powerful enough to cater for the discussion of intelligent or declarative processing.
- The concept of '**Real Systems**', introducing a modern System Theoretic understanding of the term 'system' – an understanding that is precise enough to support the deduction and analysis of numerous effects in these systems, such as autopoiesis, cognition, perception or reflection.
- The concept of a '**Microscopic-Macroscopic Dichotomy**' that provides a scientific framework for a highly prominent phenomenon: **Emergence**. A concept that later helps to understand the emergence of macroscopic effects, such as behavior, knowledge or intelligence.

4.1 Classic Systems

4.1.1 Structure

If the world is approached from the perspective of modern Physics, it is to be understood as formed by only four fundamental interactions (see Table 4.1), twelve matter constituents and their anti-particles called Fermions (see Table 4.2) and seven[16] force carrier particles, the Bosons (see Table 4.1), which are exchanged in the fundamental interactions (BARNETT, 2000 [6]).

Table 4.1. Fundamental interactions

Interaction:	Gravitational	Strong		Weak	Electromagnetic
		Fundamental	Residual		
Acts on: (Particles)	Mass-Energy (all)	Color Charge (Quarks, Gluons)	Res. Color Charge (Hadrons)	Flavor (Quarks, Leptons)	Electric Charge (electrically charged)
Mediators: - Bosons -	Graviton	Gluons	Mesons	W^+, W^-, Z^0	γ

[15] The introduced understanding of processes is new to the domain of Industrial Engineering in the sense that enterprise processes typically have been defined with an emphasis on repeatability, consistency and in particular with a well-defined goal or purpose in mind. The definition introduced here does not build on this classic framework.

[16] Among the seven force-carrier particles mentioned is the graviton, which is suspected of mediating the gravitational interaction. Its existence has not been proved yet.

Table 4.2. Matter constituents

Fermions – Matter Constituents						
Leptons	ν_e electron neutrino	e electron	ν_μ muon neutrino	μ muon	ν_τ tau neutrino	τ tau
Quarks	u up	d down	c charm	s strange	t top	b bottom

It is worth mentioning that the fundamental interactions force an immediate existence of structure in that the elementary matter constituents only exist in configurations of two (mesons) or three (baryons) bound constituents (s. Table 4.3) – a phenomenon known as quark-confinement[17]. Short- and long-range interactions drive the formation of complex structures out of compounds (protons and neutrons, both baryons) and non-compound (electrons) elements: atoms and molecules. The gravitational and electromagnetic interactions even facilitate a formation of structure over large scales in the spatial dimensions of space. All macroscopic structures that exist in the universe are for the most part a direct consequence of one or even the interplay of both of them.

Table 4.3. Subatomic structure

Hadrons – Subatomic Structures						
Mesons bosonic	Name (constituents)	π^+ pion ($u\bar{d}$)	K^- kanon ($s\bar{u}$)	ρ^+ rho ($u\bar{d}$)	B^0 B-zero ($d\bar{b}$)	... (ca. 140 types)
Baryons fermionic	Name (constituents)	p proton (uud)	n neutron (udd)	Λ lambda (uds)	Ω^- omega (sss)	... (ca. 120 types)

[17] The fundamental strong interaction shows the remarkable property that any of the particles it acts on only exist in bound states. Any experiments conducted in the field of particle physics show that all the observable particles have an integer electric charge – an integer multiple of the elementary electric charge e. Quarks with 1/3 e or 2/3 e electric charge are said to exist only in bound states (HAUBOLD and MATHAI, 1998 [36]).

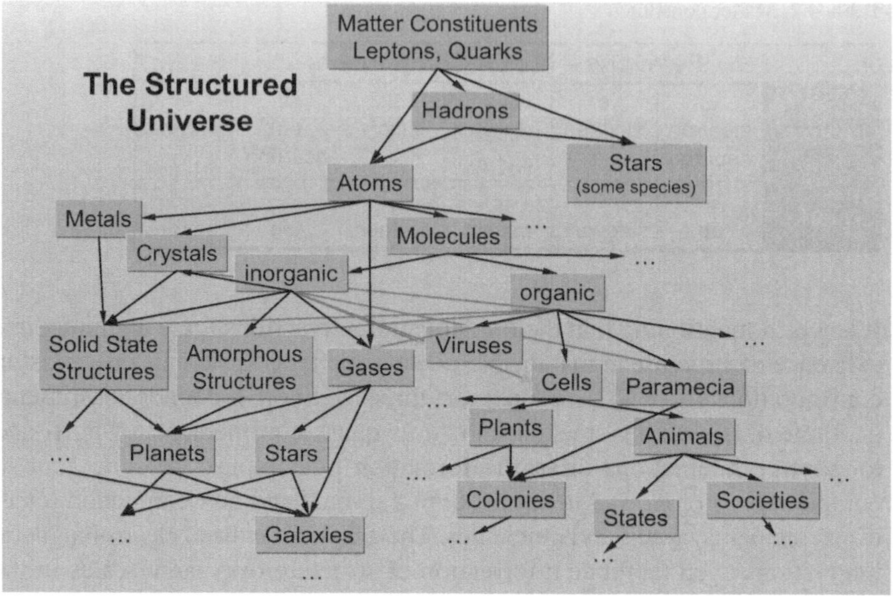

Fig. 4.1. Structures of the universe

4.1.2 Interactions and Properties

The fundamental interactions result in forces that change the particles' states. These state changes occur through the exchange, generation or extinction of force carrier particles. All interactions act on specific particle properties. In the case of subatomic particles these properties are defined by their quantum states, mass and lifetime. Neglecting interferencing effects[18], compound structures in principle inherit their properties from the constituting elements. However, interactions of compound structures are typically based on the interplay of many of their constituents interacting collectively and are thus characterized by their complexity as well as their temporal and spatial dilatation. In spite of being realized by uncountable fundamental interactions sequenced over time, interactions between complex structures are typically not perceived as a scheme of fundamental interactions, but rather as a new set of macroscopic phenomena, such as pressure, mechanical force, heat transfer, etc. This is equally true for the properties of complex compound structures that mediate macroscopic interactions. These macroscopic properties include elasticity, temperature, heat capacities and the like.

---

[18] Effects such as the mass defect (which goes hand in hand with the confinement energy in bound states) lead to slight changes in the properties of bound configurations compared to what one would expect them to inherit from their constituents.

Properties and interactions drastically change their phenomenological appearance with the scale of observation. They demonstrate the need for a microscopic-macroscopic dichotomy to understand and describe the underlying nature of numerous effects.

4.1.3 The Process

In most cases interactions are not 'happening' isolated but are effectuated by circumstances of which they are a logical consequence. Typically a single interaction is just one link in a chain of interactions, triggered by a preceding event and resulting in an event that causes follow-up interactions. In the natural sciences these sequences of interactions are called processes:

A process is a course of events and actions that are coupled by the principle of cause and effect. [Def 4.1]

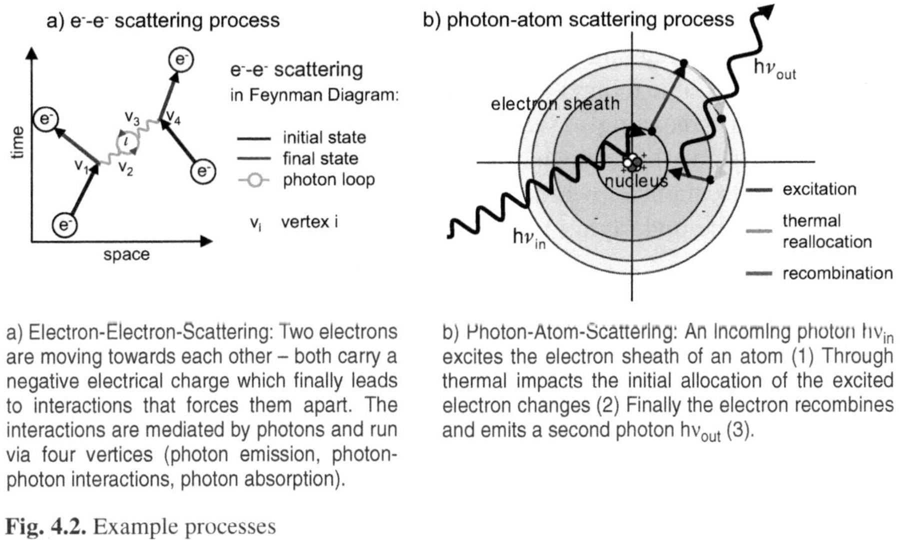

a) Electron-Electron-Scattering: Two electrons are moving towards each other – both carry a negative electrical charge which finally leads to interactions that forces them apart. The interactions are mediated by photons and run via four vertices (photon emission, photon-photon interactions, photon absorption).

b) Photon-Atom-Scattering: An incoming photon $h\nu_{in}$ excites the electron sheath of an atom (1) Through thermal impacts the initial allocation of the excited electron changes (2) Finally the electron recombines and emits a second photon $h\nu_{out}$ (3).

Fig. 4.2. Example processes

In simple interactions, event and action tend to be merged together into a single vertex, whereas high-level interactions are typically characterized by a separation between the causing events and the triggered actions. This temporal and spatial separation is what causes the macroscopic perception of complex structures to behave as a 'sensor-actuator system', which encapsulates an arbitrary and often invisible set of inner processes.

Philosophy distinguishes these so-called 'causal' processes from 'pseudo' processes. The latter lack any form of causation between events and actions – thus they will not be discussed any further in this work. The causal processes are identified by what Russell introduced as 'causal lines', defining a more general and abstract form of causation:

> *"I call a series of events a causal line if, given some of them, we can infer something about the others without having to know anything about the environment." (RUSSEL, 1948 [88], p. 333)*

The (causal) process implies an underlying principle or coordinating logic leading to responsive behavior in contrast to random or uncorrelated behavior. In other words, it is the process that guarantees a behavior that is responsive to the triggering event. In addition, the process inherits a notion of predictability from its underlying coordinating principles[19]. Responsive behavior thus is characterized by exactly this notion of predictability.

Russell's introduction of 'causal lines' also provides some hints to what philosophers call 'Meaning'. Given some of the events of a process, the observer is only able to infer something about the other events due to the fact that they form a meaningful sequence. The first events have to have meaning for the observer in the sense that she/he can perceive their causal relationship and their consequence towards follow-up events. Meaning for the observer thus reflects the underlying principle or coordinating logic of an observed process.

In this work, Def 4.1 was adopted as a general definition of 'process'. It is worth mentioning that this definition significantly differs from the general understanding of 'process' in the business and engineering worlds, where processes are typically defined with respect to a certain goal or desired effect and with a strong emphasis on consistency and repeatability. Definition 4.1 should be general enough to cater for the radical new understanding of 'process' demanded in Chapter 3.2.

4.1.4 The System – A Classic Definition

Classical Philosophy invented numerous analytical methods to better discuss and understand all the structures surrounding us. One of these methods was

[19] The predictability is even given for the process examples illustrated in Chapter 4.1.3, Fig. 4.2, although quantum mechanics implies that their exact outcome is subject to a certain randomness. It is still predictable in the sense that the outcome will follow a well-defined probability distribution.

'System Thinking': describing the world in terms of systems and their environments facilitates the perception of patterns, rules and relations – an approach later adopted by many scientific disciplines.

Pioneering modern Thermodynamics, Gibbs defined:

> *"Any portion of the material universe which we choose to separate in thought from the rest of the universe for the purpose of considering and discussing the various changes that may occur within it under various conditions is called a system." (RUKEYSER, 1942 [87], p. 445)*

The system in this sense is a mind construct formed by drawing a boundary that separates the system from its environment. The boundaries – encapsulating structural components, interactions and processes in a system – are typically drawn in a way that simplifies the analysis and description of

- interactions with the environment (I/O relations)
- internal interactions and processes as well as
- the sequence of states the system traverses.

System Theory discusses systems in terms of their 'Structure' and 'Organization' as the two constituting elements that form any system (structure-organization-dichotomy) (MATURANA, 1975 [60]; WHITAKER, 1995a [117]).

The '**Structure**' (Latin: structura – a fitting together, building; Latin: structum – something that has been built) is the set of physical components in a given space. It determines the space in which the system exists and can be perturbed.

The '**Organization**' (Greek: organikos – serving as instruments, instrumental) is the instrumental participation of the components in the constitution of the unit. It defines the components' functional interplay and all possible processes that the system can enact.

It is the system's organization that determines the way it interacts with its environment and thus defines the system's overall character as perceived by an observer. The system's identity thus is bound to its organization!

4.2 Real Systems

Throughout the history of the universe uncountable structural configurations of varying size and complexity formed and defined the character of their era. Most of them, however, had properties that dictated or demanded necessary conditions[20] under which they could persist. With the conditions drastically changing over time, almost all of them disappeared again. What is left over forms the universe as it is observed today.

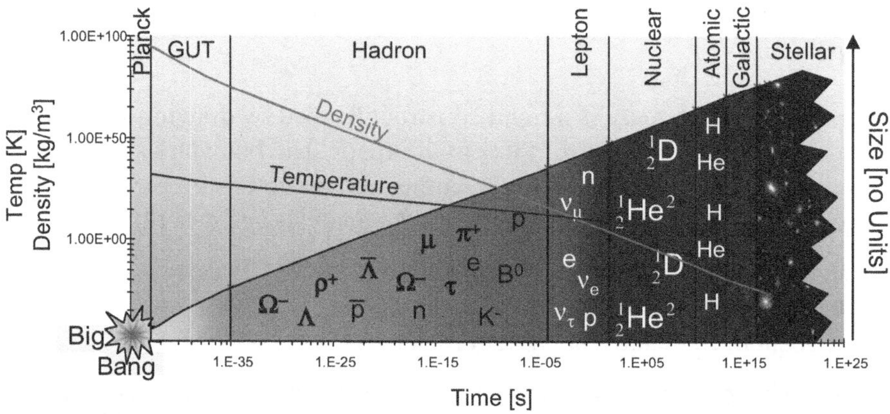

Next to the temperature and density development over time Fig. 4.3 shows the epochs that are commonly distinguished in astrophysics. Each of the epochs (except the first two and the forth) is named by the structures that have formed and dominated their time. Most of these structures (e.g. heavy hadrons, or in particular any anti-mater structures) have disappeared (or exist only in extremely small quantities) during the development of the universe[21].

Fig. 4.3. The development of structures in the universe

In a sense, structure itself is exposed to an evolutional development. All configurations that are observed have to be of a certain rigidity allowing them to persist long enough to be observed a significant time after their integration. Exposed to a constant decay, the structures that naturally exist today

a) had to be integrated once in extremely large quantities or
b) have to be constantly produced/reproduced or
c) have to reproduce themselves.

[20] These conditions should include lifetime considerations in the sense that implicitly they demand to be observed within their lifetime! Other conditions include thermal circumstances, chemical environment, etc.

[21] Data to build Fig. 4.3 was taken from Chaisson, E. and McMillan, S.: Astronomy Today 3rd Edition, chapter 27. (CHAISSON and MCMILLAN, 1998 [16])

Due to favorable conditions numerous such c)-type reproducing structures are observed on earth (Biology). However, all of them are less characterized by their assemblage of structural components than by the ongoing processes they enact. Even their existence relies purely on the fine-tuned interplay of components that constantly process energy and building blocks to preserve the existing order and resists the 'entropic' decay. Analyzing them in depth, one indeed has to admit that the structural components play a minor role and are even constantly replaced, built up and removed. They only provide the physical infrastructure, the space or the pegs, to which the processes hang on. Being characterized by the set of processes and the organizational setup defining the interplay of the constituting components rather than by the components themselves, these reproducing structures are in most cases phenomenologically better understood as systems than as sole structures: they are biological systems.

A necessary requirement for the ability to reproduce as a whole is the ability to reproduce structure and organization. The biologists Maturana and Varela indeed identified the ongoing reproduction of structure and organization as fundamental to all living systems. They argue that this requires the alignment of all inner processes to self-production and self-regeneration. Only an excess capacity in this inner production of structure and organization in the end facilitates the reproduction of the system as a whole.

In 1972 Maturana coined the term "Autopoiesis" combining 'auto' (Greek: self-) and 'poiesis' (Greek: creation, production) to name the phenomenon of inner self-reproduction. It should later become the central concept of System Theory (REESE-SCHAEFER, 1992 [81]), where it is commonly defined as:

Autopoiesis[22] is the ability of a system to generate its specific constitution – its components (structure) and their interplay (organization) – on its own. [Def 4.2]

[22] The formal definition of 'Autopoiesis' originally introduced by the biologists Maturana and Varela is given by: "An autopoietic system is organized (defined as a unity) as a network of processes of production (transformation and destruction) of components that produces the components that: 1. through their interactions and transformations continuously regenerate and realize the network of processes (relations) that produced them; and 2. constitute it (the machine) as a concrete unity in the space in which they [the components] exist by specifying the topological domain of its realization as such a network." (VARELA, 1979, [112], p. 13)

With their defining characteristic – the ability to reproduce – being the only reason for their existence and with 'Autopoiesis' as a clear functional definition for the realization of their characteristic – that requires numerous functionally inseparable processes and physical components – living systems represent a special type of systems: **Real Systems**[23]. It is not anymore up to the observer what to regard as part of the system and what not because 'Autopoiesis' can only be achieved by the unity of the components and their specific organization. They are not mind constructs as their existence is evident. As they exist, they realize their own character and thus define their own system border, which encapsulates all those structural components and their interplay needed to realize this character.

The fact alone that autopoietic systems are real systems indicates a sharp turn away from the classic understanding of systems, where they were defined as mind constructs defined by the observer by drawing a system border that encapsulated whatever was thought to be useful. It is the dawn of modern 'System Theory': **real systems exist**[24]!

Luhmann generalized the understanding of the autopoietic phenomenon and transferred it to the level of abstract systems. Based on this abstraction he developed a general theory of social systems (LUHMANN, 1984 [53], 1986 [54]).

Applications of Luhmann's approach include a theoretical description of legal systems (TEUBNER, 1988 [107]; TEUBNER and FEBBRAJO, 1992, [108]), the field of accounting (ROBB, 1991 [82]) or the family (ZELENY and HUFFORD, 1992 [123]).

Real systems in Luhmann's sense do not necessarily have to be living (Luhmann, 1986 [54]). Rather, there are numerous nonliving systems that can be regarded as autopoietic and thus real. Even the systems' structural elements do not necessarily need to be of a physical nature, as long as they show properties for interactions to act on and mediate between process steps in the sense that they post follow-up events or facilitate the change of system states at a triggering event.

---

23 Real systems in the sense that they are not fictitious mind constructs. Real Systems are systems that mark out their own system border, not leaving it up to the observer what to regard as part of the system and what not.

24 Note that the existence of real systems as introduced here does not necessarily call the constructivistic world model into question. A world in which systems exist that mark out their own system borders might still be nothing else but a mind product.

4.3 Fundamentals of Autopoietic Systems

The autopoietic system is characterized by its organization which manifests in the set of processes the system enacts at a given time t. By definition 4.1, the process was introduced as a course of events and actions coupled by the principle of cause and effect (event-action-world). Embedded in the conceptual framework 'System', however, a new perspective seems reasonable:

At any instance of time t, a system could be described by defining all its constituents, their properties or attributes and their relation to each other. Given the case that all these parameters could be captured by a set of variables v_1, v_2, ..., v_n ($n \in \mathbb{N}$), Cybernetics would describe the system at time t_0 as being in a particular state s with all variables showing discrete values $v_i = v_i(t=t_0)$.

> *"The state of a system at a given instant is the set of numerical values which its variables have at that instant." (ASHBY, 1960 [4], p.16)*

With every action changing the condition of one or more structural elements, their attributes or relations to each other, it is valid to redefine the process in the state world:

A process is the sequence of states a system passes through with each transition following an underlying principle or guiding logic. (State world) [Def 4.3]

System Theory identifies meaningful processing as a fundamental function of every system (REESE-SCHAEFER, 1992 [81]). Processing in the sense of system theory is the sequential selection of activities forcing the system into new states (state transitions). The initial trigger for processing is information, which is regarded as an event that selects a system state.

As an autopoietic system constantly replaces its constituents and redefines their organization, every system state leaves a structural effect and every state transition leads to a new state exclusively derived out of the previous one. Irreversibility thus is an inherent property of all autopoietic systems – **history never repeats but it forms the future**!

Macroscopically, the system appears to select a follow-up action or a new state when perturbed by any stimuli. The selections carrying the process from state to state are said to be the fundamental operations of an autopoietic system. Processing is meaningful if these selections are meaningful. Meaning is the functional control of the selections. Meaning thus was stipulated as a fundamental resource (REESE-SCHAEFER, 1992 [81]). However, if we recall the definition of 'process', the selections are all of a rather passive nature – they just happen – as they are defined by the underlying coordinating principle. In physical systems meaning thus is represented by the consequences of the 'principle of cause and effect', manifested by the physical laws to which matter obeys. Constantly changing structure and organization, these laws don't change, but their consequences do! Meaning thus develops along with the organization in the processing autopoietic system. Moreover, meaning builds on the system's organization and thus is engraved physically. With the underlying physical laws not being deterministic (but rather defining their consequences in terms of probability distributions when it comes to describing microscopic structures[25]), meaning shows yet another characteristic: it is not necessarily strictly deterministic but leaves room for a certain randomness.

Autopoietic systems show a remarkable property in the way they interact with their environment: on the one hand building blocks and energy (including information) are exchanged with the environment, which characterizes them as open systems. On the other hand any functional mechanisms, the way the system processes, incorporates building blocks and responds to information are totally self-determined and cannot be controlled by interventions from the environment. "While a given perturbation may trigger a change of system states, the particular change triggered is a function of the system's own organization and structure" (WHITAKER, 1995a [117], p. 5). Even the way the system develops its meaning is self-referential, building purely on its own organizational setup[26]. The autopoietic system thus is functionally closed.

4.4 The Microscopic-Macroscopic Dichotomy

System Theory discusses systems from a macroscopic perspective. It describes the systems' macroscopic behavior, functional mechanisms and how processing is realized by meaningful selections.

---

25 A direct consequence of the uncertainty principle of Quantum Mechanics.

26 As meaning is built up totally self-referentially learning has to take a similar self-referential way: In principle, learning may happen through irritations or disappointed expectations. As perception itself is again guided by the system of meaning, any evaluation schemes available and any sensors for irritations lose their objectivity! Nevertheless, this mechanism in principle allows an 'external reference' to indirectly stimulate learning processes.

The previous introduction to real systems was driven by a focus on microscopic phenomena, introducing the process, structure and organization microscopically as based on matter constituents, the principle of cause and effect and the fundamental physical interactions.

The phenomenological difference between the appearance of effects, interactions, mechanisms and properties in the two perspectives exemplifies the microscopic-macroscopic dichotomy resulting from the transition from the individual and simple to the collective and complex.

Just as the kinetic energy of microscopic constituents is macroscopically perceived as the phenomenon heat or the electrodynamic interactions of the constituents of a solid body are phenomenologically identified as mechanical forces, so are many of a system's microscopic phenomena subject to a shift in appearance once they are observed from a macroscopic perspective (**Emergence of Macroscopic Phenomena**). The understanding of the microscopic origin of a macroscopic phenomenon is in most cases crucial for their scientific understanding and discussion.

Information as introduced by System Theory is the collective effect brought about by numerous microscopic events perturbing the system. Biologically it is a stimuli pattern that suddenly occurs or changes – resulting in a perceivable difference, the prominent basic element of cognition.

For the natural sciences **meaning** classically was a highly intangible and complex philosophic concept. By tradition it was understood as the sense, significance or value something had to the observer. Only since Cassirer, Durkheim, and Parsons have many 'substance terms' been resolved in 'function terms' (functionalism) (REESE-SCHAEFER, 1992 [81]) and meaning was slowly redefined functionally. As introduced here, meaning is a macroscopic effect that results from the microscopic organization of the system. It is the effect of all involved constituents collectively enacting the fundamental laws that manifest the coordinating principle's defining the system's network of processes. Meaning as the consequence of the principle of cause and effect is thus the control of the process flow. With System Theory introducing processing conceptually as a sequential selection, the selection, too, has to be understood macroscopically! The selection conceptualizes the meaningful transition from one macroscopic state to another, in between which microscopically uncountable intermediate states may be passed through unnoticeably for an outside observer. The selection thus is a fragment of a larger process, a procedure or simply a process itself which is controlled or even defined by meaning. To sum up, **selections** are the macroscopic manifestation of microscopic processes that enact the transitions between microscopic states, whereas **meaning** is the macroscopic manifestation of microscopic effects (laws and principles) that direct these transitions.

System behavior is another often discussed macroscopic phenomenon. Behavior conceptualizes the sum of interactions a system has with its environment. These interactions are enacted by system processes. Behavior thus is the macroscopic manifestation of microscopic processes.

The understanding of the microscopic-macroscopic dichotomy allows macroscopic systems to be understood microscopically and vice versa. With a focus on technical and functional aspects it allows to grasp and describe complex macroscopic systems by a simple microscopic representation: a sensor-actuator system. Unique to all autopoietic systems is their underlying interaction scheme, which begins with information (stimuli) that triggers the assembly of a response process (controlled by the system's meaning) and ends with the enactment of whatever activities are trigged (actuators) as a result of this assembly.

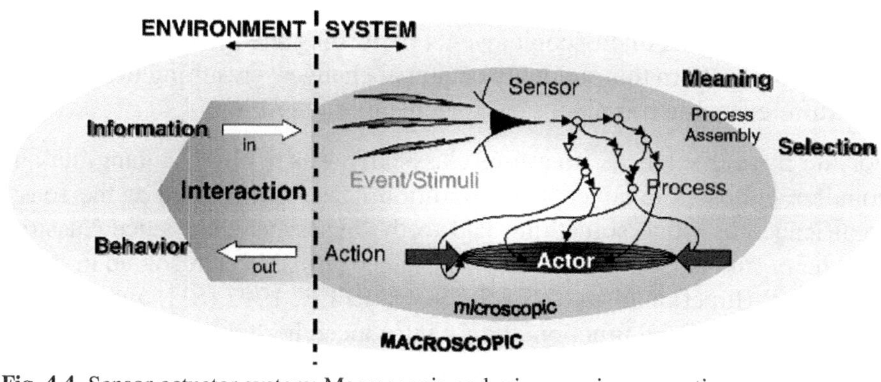

Fig. 4.4. Sensor-actuator system: Macroscopic and microscopic perspective

4.5 Social Systems – Communication as a Structural Element

Any approach that discusses social systems, such as organizations, communities, enterprises and the like, based on System Theory leads to the fundamental question: What is the role the individuals play?

Analyzed more closely, these individuals

- are identified as autopoietic systems themselves.
- implement their processes based on their own structural and organizational setup – thus they cannot be functionally controlled by the social system but can only be motivated to act in its sense.
- are not necessarily reproduced in the system, and if so, they reproduce themselves rather than the social system reproduces them.
- are members of several communities, societies and organizations at a time, withdraw from them and join them mostly of their own free will.

Except the last point this is a situation that was identically discussed already in chapter 3.2. Luhmann consequently argues that the individuals are not the structural components of social systems. He rather suggests that communications should be regarded as the constituting structural elements of social systems (LUHMANN, 1986 [54]).

Conceptualizing them as structural components makes communications a highly complex and rather controversially discussed phenomenon. Communication in this sense cannot be regarded as a kind of action, as this would not provide enough conceptual power – even though communications and actions have to go hand-in-hand. Communications have to interact with each other – they have to show properties for interactions to act on. There does exist something like a minimum size, and communications have to have a temporal persistence although they may or even have to decay over time (LUHMANN, 1986 [54]).

Luhmann's concept of communications was heavily discussed and criticized (WHITAKER, 1995 [119]). His suggestion that they be regarded as a synthesis of three selections – information, utterance and understanding – do not seem to be sufficient to define structural components or the space they define for the system to live in. Nevertheless, they gave rise to uncountable discussions on the role of communications, language and languaging. A concluding discussion on the concept of communications and their nature is given in chapter 7.6.

4.6 Cognition, Reflection and Memory

System Theory made some of the most significant contributions by providing a new understanding of reflection and cognition, both of which have been isolated from human consciousness and the human brain in that they have been transferred to the abstract system level. This provides a starting point to discuss intelligent systems, learning and the management of system knowledge – as introduced in chapter 5.

4.6.1 Cognition

With the beginning of the 20th century the "Thinking in Differences" or "Philosophy of Differences" gained significant prominence, being discussed by philosophers like Saussure, Heidegger, Deleuze, Derrida and Luhmann (SAUSSURE 1916/1975 [90], HEIDEGGER 1957 [37], DELEUZE 1997 [21], DERRIDA 1986 [22], REESE-SCHAEFER, 1992 [81]). From then onwards 'difference' was understood as the basic element of cognition and 'distinction' as its fundamental operation. A cognitive system "... discriminates among differences in its environment and potential states of that environment" (WHITAKER, 1995a [117], p. 7), enabling successful behavior in response to dynamic changes.

From the biological point of view, cognition is the ability of a living system to continuously compensate perturbations.

Here, 'perturbations' are sudden stimuli patterns or changes in these patterns. The 'compensation' is achieved by a reasonable response that ensures the system's existence and enables its self-maintenance.

In a similar sense the cognitive domain (cognitive reality) is identified by Maturana and Varela in 1980 as "... all the interactions in which an autopoietic system can enter without loss of identity. This implies to regard cognition as the active (inductive) acting or behaving in this domain." (MATURANA and VARELA, 1980, [58] p. 13)

Given the previous discussion (Chap. 4.4), any interaction consists of the triggering event (information, stimuli), the meaningful assembly of a response process and the enactment of activities. The system's identity is preserved as long as the incoming event does not trigger the assembly of a process that violates the system's meaning. Microscopically the assembly of a process that violates the system's meaning is simply not possible (meaning reflects the consequences of stimulus propagation!). Such a stimulus will simply not lead to the successful assembly of a response process.

The cognitive domain thus is given by all those interactions (a system can enter) for which the system can successfully assemble – guided by its meaning – a process as response to the triggering event. [Def 4.4]

Cognition is the actual 'acting' or 'behaving' within the cognitive domain[27]. [Def 4.5]

A system is said to have successfully perceived a stimulus if and only if it has been able to successfully assemble a response process based on its organizational setup (meaning). [Def 4.6]

These definitions introduce cognition and perception on the level of abstract systems, phenomenologically liberated from their classic 'boundness' to the human mind. More than this they imply every autopoietic system to be a cognitive system, as it naturally does exactly what the definition requires: meaningful processing.

4.6.2 Reflection

Reflection, or the process of thinking as it is understood classically, is the generation of new meaning out of the 'System of Meaning' that one has already built up.

Autopoietic systems self-referentially generating and regenerating their organization do exactly this! On the one hand meaning controls the assembly of processes, on the other hand these processes form and generate structure and organization and thus new meaning, which is microscopically defined through the organizational setup of the system. The system of meaning itself thus has an autopoietic character. Autopoiesis of the system of meaning is another word for reflection, which is now no longer to be understood as a phenomenon of the human domain (REESE-SCHAEFER, 1992 [81]).

4.6.3 Memory

Since the biologist Humberto Maturana identified that there is no physiological evidence for memory in natural brains the role, necessity or even existence of memory has become a central issue in System Theory (REESE-SCHAEFER, 1992 [81]).

Autopoietic systems constantly reproduce their organization based on their organization. History thus plays a fundamental role, but is functionally inherent in the system in the form of its organization. There is no need for a registry of bits and bytes, for files, data stores or even stored relations, sentences and expressions that would represent a model of the world. By their

---

[27] A definition adopted from Maturana and Varela (MATURANA and VARELA, 1980, p. 13).

very nature autopoietic systems incorporate behavior or reaction patterns. Only if observed from the outside, the effect that similar stimulus patterns trigger similar response behaviors may lead the observer to attribute memory to the system.

Even if memory would exist, its use would require a highly developed 'System of Meaning' and highest intelligence, as

- the development of a representation language and its full incorporation has to be realized,
- the fully conscious and intentional use of this language has to be managed,
- the ability to process and rebuild content out of descriptions would have to be built up.

Memory like representations, reflecting the world, world models, high-level languages, etc. may be a golden goal for any intelligent system but never a starting point!

Chapter 5 Theory of System Knowledge

An Introduction to the Proposed Theoretical Framework – Understanding and Describing Knowledge and Intelligence on an Abstract System Level

As System Theory has phenomenologically introduced cognition, perception and reflection on the level of abstract systems, there now exists a sound scientific basis for discussing knowledge and intelligence isolated from the human mind and generalized in the same manner to the level of autopoietic systems.

Whether the System Theorists avoided the direct contact or association with Epistemology – a very traditional, highly developed but also slightly stagnating discipline[28] – or knowledge and intelligence simply have not been of high interest for them, both did not earn prominence in their discussions. Luhmann once described knowledge as redundancies that accelerate processing (LUHMANN, 1997 [51], p. 124) and another time as socially accepted and generalized solutions to known problems in the sense that knowledge is what the social system defines as knowledge (LUHMANN, 1992 [52], pp. 107).

Continuing the conceptual approach from chapters 3 and 4, knowledge and intelligence should be introduced phenomenologically from the perspective of the natural sciences, in order to cater for the needs of business and engineering management.

5.1 Intelligent Systems

The nature of intelligence, how it is best measured and scientifically discussed, is subject to endless debates. In an effort to search for a suitable definition the American Psychological Association (APA) in 1995 brought together no less than two dozen prominent theorists who should bring clarity to this issue (APA, 1995 [2]), but still there exists no generally accepted agreement.

---

[28] See the discussion in Reese-Schaefer's introduction to Luhmann's theories (REESE-SCHAEFER, 1992 [81], Chap. 3)

In the common use of language, intelligence is most often associated with abilities such as practical problem solving (involving logical reasoning, seeing all sides of a problem and keeping an open mind), verbal ability (being a good conversationalist, reading well or often) and social abilities (admitting mistakes, displaying interests in the world at large) (RADLER, 1999 [79]). Intelligence is often measured by psychometric factor analyses that provide a single unified result, such as Spearman's g-factor, or a number of factors rating different mental abilities separately. These explicit measures are still regarded as fundamental by many researchers (APA, 1995 [2]), although it seems evident that intelligence cannot be separated from numerous implicit effects that are not taken into account by most psychometric approaches (WEINBERG, 1989 [116]; APA, 1995 [2]).

Findings that simple perceptual and cognitive tasks are performed at greater speeds or the ability of adaptation to unfamiliar tasks and life situations is higher with intelligent persons may provide more hints to a phenomenological understanding of intelligence.

In his 'Triachic Theory' of intelligence Sternberg offers a phenomenological approach that distinguishes three processes to specify the nature of intelligence in terms of behavioral schemes and relationships between the internal and external world (STERNBERG, 1988 [97]). Analyzing human intelligence with a focus on how it is gained, built up and developed, the psychologist and epistemologist Piaget identified adaptation as a key to the phenomenological understanding of intelligence and its effects (PIAGET, 1972 [71]).

Adaptation is probably the most fundamental phenomenon identified to bring about intelligence and in most cases the fact that it plays a central role is the smallest common denominator among the many scientists discussing this issue. The incorporation of behavior patterns is a clear act of adaptation, and so is the adoption of methodologies, concepts, approaches, logical schemes etc. that finally lead to the ability of high-level reasoning and logical deduction. Adaptation is the microscopic mechanism behind any kind of learning and higher-level acquisition of knowledge. Transferring intelligence to the level of abstract systems, adaptation is phenomenologically simple enough to cater for this abstraction: a system that adapts well simply minimizes irritations in its interactions with the environment and behaves effectively in it. In the simplest case a system may adapt to different stimuli by enacting different predefined processes as a response. If, however, the system's perceptive ability to distinguish between events does not provide the necessary differentiation, or the system does not have any appropriate response processes available, it is unable to adapt any further. Adaptation to sustaining environmental changes can, and eventually will, only occur[29] through natural selection and extinction of individual systems, leading to an adaptation of the population but not the individual (Evolutional Learning, see Chap. 3.2). High-level systems, however, are able to adapt the processes themselves that are enacted as a response to a given type of trigger. They thus are able to adapt within their own lifetime and with it fulfill the most elementary requirements for learning and the incorporation of behavioral patterns. This in turn clearly qualifies them to be regarded as intelligent (Intelligent Learning, see Chap. 3.2).

Given the System Theoretic background discussed in the last chapter, adaptation or any kind of adaptivity finds its origin in the autopoietic system's ability to constantly regenerate and realign structure and organization. Only this physical feature enables the refinement of processes and behavior patterns by changing the organizational setup that defines the interplay of a system's structural components.

In addition to this fundamental physical change mechanism, adaptation needs an evaluation scheme to drive changes into the right direction. As discussed earlier, such schemes can only be built up self-referentially – the system perceives its success or failure by evaluating environmental stimuli after enacted behavioral patterns; however, this perception purely depends on its own 'System of Meaning' – but they are not totally decoupled from the system's real behavioral effectiveness as ongoing irritations or disappointed expectations eventually will enforce refinements[30]!

As natural organisms tend to supply the enacting process chains with an excess of resources, energy and building blocks, it is valid to assume that any

---

[29] Given that the individual systems reproduce by passing on heredity from generation to generation and given that the reproduction generates sufficient variety among the individuals, the population can be assumed to be adaptive to environmental changes through natural selection and extinction.

[30] Piaget describes the self-referential process of knowledge build-up in the human brain as 'Assimilation' that enables the integration of new elements into a cognitive framework. This framework grows with assimilation and represents the fundament of perception and behavior. Nevertheless, ongoing irritations or a lack of success in the person's behavior modify the assimilation schemes – a process Piaget calls 'Accommodation' (PIAGET, 1983 [70], pp. 32). Abstracted to the level of social systems, Luhmann, too, identifies 'self-referentialism' as fundamental to the generation of knowledge via two processes which he names 'Condensation' and 'Generalized Confirmation'. System irritations increase the possibilities of resisting communications that finally control the self-referential circle. In his sense, communication is kept in check by (resisting or opposing) communication. (LUHMANN, 1984 [53], pp. 648)

enacted process pattern is reinforced by redundant infrastructure and links – an effect that is necessary to stabilize the fuzziness of microscopic mechanisms discussed in chapter 4.3. The reinforcement may depend on the overall excitation of the system towards the triggered process or the availability of resources and energy. Obviously this reinforcement has to be kept in check by suitable evaluation mechanisms. Irritations and disappointments do have to lead to the depreciation of process patterns, to the build-up or enforcement of alternatives and eventually to the stop of future enactments of unsuitable processes even in situations where they might already have been triggered[31].

In other words, intelligent systems populate behavior patterns or process fragments by reinforcement and control these populations by selection and extinction via their evaluation mechanisms. Microscopically they thus implement a virtual evolution to direct learning and the acquisition of knowledge – a fact that has been widely recognized throughout the scientific community: the similarity of functional principles in evolutional learning and intelligent learning (POPPER, 1972 [75], 1984 [74]; LORENZ, 1977 [50]; CAMPBELL, 1974 [15]).

Summarizing the discussion, we identify intelligence as a methodological framework for continuous adaptation and the incorporation of behavioral patterns. This is realized by a virtual evolution of pattern fragments that are populated and depreciated by the system's own evaluation schemes. Highly intelligent systems are characterized by comprehensive abilities to evaluate and effectuate changes, but also by the complexity of incorporable patterns and the speed at which changes take place. The ongoing autopoietic processes reproducing structure and organization physically realize the whole methodological framework.

Consequently, processes in intelligent systems have developed to one-time instances that are constantly adapted to the perceived situation. For each and every given challenge highly intelligent systems enact a unique process that is specifically tailored to the given situation. Macroscopically this response process has to be assembled – or chained 'on-the-fly' – considering, and as a response to, all given stimuli in the past and present. This does strongly dif-

[31] Already Popper argues that knowledge cannot be 'verified' but can only be 'falsified' by experiencing counter examples. Anything that has not been falsified yet does have the epistemological status of a 'conjecture' and is to be regarded as a valid belief. A falsified conjecture is depreciated or replaced by a new one – the only mechanism that controls the population of conjectures. (POPPER, 1963 [73])

fer from the traditional understanding of 'Process' in engineering and management, where typically repeatability is emphasized!

5.2 Knowledge

The classical philosophic tradition demands for the separation of 'opinion' and the noble 'Knowledge'. Since Socrates and Plato knowledge was identified as justified true belief[32] (CORNMAN, et. al, 1987 [18], p. 43; PLATON, 1958 [72], 201b7-c7) – an understanding that was shared among most philosophers even up to the beginning of modern Philosophy.

In the 17th century the Positivists Bacon and Hobbes saw knowledge under a new perspective: 'Power'. Later the American Pragmatists, Pierce, James and Dewey, identified knowledge as a tool to cope with reality – Plato's fundamental truth, as the main criteria to identify knowledge, was replaced by the sheer success achieved (RORTY, 1982 [86]). Piaget approaches knowledge from a similar perspective, as an actively constructed framework (built up through the subject's actions and interactions) that enables successful behavior (PIAGET, 1983 [70], pp. 25). Knowledge thus drives behavior! It defines the way a system acts in its environment, thus it has more in common with meaning as discussed in chapter 4 than with facts, expressions or logical sentences that are kept in memory and are subject to (scientific) verification or falsification[33]. But is it the system's meaning alone that determines its behavior? Meaning as introduced earlier is the macroscopic control of state transitions or, in other words, the control of the selections that force the system into new states. The selections without doubt have a great influence on the system's overall behavior, but phenomenologically the elements to be selected are at least equally important.

---

[32] Up to today philosophic knowledge is still subject to scientific discussions. In particular the question of whether there is a minimal knowledge term, which was proposed by Kutschera (KUTSCHERA, 1982 [47]), or whether knowledge is merely true belief (SARTWELL, 1992 [89]) are issues of controversy in modern epistemology.

[33] 'System Knowledge' and 'Philosophic Knowledge' thus have to be regarded as two distinct phenomena. Philosophically the claim, or at least the effort, does and should exist to exploit a fundamental truth that has been driving all scientific disciplines alike. But this scientific knowledge is different from the framework based on which intelligent systems enact their behavior. This is a fact that was ignored by many that criticized the pragmatic approach, like: Rorty (RORTY, 1982 [86]). Without doubt both phenomena are interlinked as on the one hand scientific knowledge is formulated by individuals and on the other hand it allows the individuals to adopt methodologies that they did not need to develop and perceive themselves – this would require a deeper investigation and should not be discussed here.

With 'Behavior', 'Selection', 'Transitions', but also 'Power' or 'Tools to Cope with Reality' being terms typically devoted to the macroscopic domain, this short introduction already indicates that knowledge is almost exclusively discussed from a macroscopic perspective. However, the phenomenological understanding in both, macroscopic and microscopic, domains is of high importance to the design and implementation of any efforts to manage knowledge. This is in particular the case as we do have to expect a great change in the phenomenological appearance of knowledge with the scale of observation (see Chapter 3.5). Both perspectives are to be discussed in the next two sections.

5.3 Knowledge from a Microscopic Perspective

Scaling down to the lowest system levels, such as organic molecules or even atoms, we find that their behavior is simply given by the set of physical processes they can enact on perturbations. Even if regarding the simplest living system, a few levels up, we observe that physical and chemical processes, which are engraved in the system's structural setup, purely define their behavior. On even higher system levels, microscopically it is still biological processes, motional sequences, social processes etc. that are in control of the system's behavior.

As mentioned in chapter 3.1, integrations of higher-level systems occur primarily as they provide an evolutional advantage to the participants. Functionally this advantage is gained by new degrees of freedom that offer new possibilities for a better overall adaptation and a higher effectiveness. Increased degrees of freedom, however, always come with a higher overall complexity that needs to be managed in order to turn potential abilities into real capabilities.

The identified processes exactly provide this management. Their underlying principles or logic constrains and coordinates the use of these new degrees of freedom gained through the system's integration. They allow the system, as a whole, to adapt better to the environment than it would be possible for the participants as sole players.

It is important to notice that any degree of freedom already brought in by the participants is in general out of the system's scope. The participants of higher-level systems are themselves operationally closed and self-determined units that – as Luhmann proposed – cannot be regarded as the system's structural components but rather as its environment. The participants fully control all those degrees of freedom that they bring into the higher-level system – they are to be addressed with a notion of 'What to Achieve' rather than 'How to Achieve it' and will enact whatever is necessary to fulfill the request. As the higher-level system in a sense 'ends' at the system border of the lower-level participant, their degrees of freedom can clearly be regarded as 'out of the system's scope'!

To illustrate this, a single muscle cell (participant) in an animal body (system) shall be regarded as an example. The contraction of the cell and how it is enacted is out of the system's scope – it is enacted by the muscle cells, given the appropriate stimulus that provides the cell with a notion of 'What to Achieve' ("contract now!") but not how to achieve it. The coordination of thousands of muscle cells, however, that leads to a macroscopic movement of the animal's body parts – a new degree of freedom gained only through the integration of the system – is a typical biological process which defines the system's behavior.

Note further that on the one hand it is only these 'pre-controlled' and 'ready to execute' activities of the participants that facilitate the system's function and behavior at all, on the other hand, however, does the behavior of the participants only weakly influence the system's macroscopic behavior!

Taken as a further example, some muscle cells (participants) in a human body (system), it is again their proper coordination that may lead the person's head to shake from left to right. If the participants improve their performance, precision, response time or contracted length, this may lead to a more intense, more precise, faster, etc. shaking from left to right, which is more or less the same system behavior – even if all of the participants collectively improve their activities! A significant change of behavior requires changes in the coordination of the involved muscle cells and thus the enactment of a different process. Only now the person's head may shake up and down – again depending only little on the performance of the muscle cells.

The situation may look different as soon as physical performance issues play a significant role in characterizing the system's overall behavior. As intelligent systems, however, tend to operate well below their critical performance marks, it can be regarded as a good approximation to describe the behavior of intelligent systems[34] as dependent only on the coordinating processes!

Given that knowledge is the underpinning framework of a system's behavior, as introduced in 3.2, the discussion above can be concluded in the following definition of knowledge from a microscopic perspective:

System knowledge under a *microscopic* perspective is given by those incorporated processes that coordinate and constrain the use of system-specific degrees of freedom gained through system integration.

[Def 5.1]

Contrary to the macroscopic perspective introduced in the next paragraph, this definition provides an understanding of the implementation layer. It describes how the system implements its behavior: by the enactment of microscopic processes that coordinate the system's degrees of freedom.

5.4 Knowledge from a Macroscopic Perspective

The macroscopic perspective provides a view on systems that is characterized by a focus on their macroscopic behavior, how it is achieved, its overall expressive notion and intention. If we take the second example in 5.3, the shaking head macroscopically indicates a notion of acceptance or rejection (yes or no), which may be expressed in many different ways. Macroscopically it is less interesting how the coordination of muscle cells finally leads to a shaking head rather than questions like: 'What triggered this response?', 'How was the reasoning carried out?' or 'Why has the head-shake and not a verbal communication been enacted?'. Obviously a clear line of demarcation cannot be drawn between macroscopic and microscopic perspective – this is why it is even more important to understand both views.

As discussed in chapter 3.1, in intelligent systems the process does not exist as a predefined entity. Rather, a specific process is instantiated only once for a given challenge in its specific context. This requires mechanisms to assemble and enact processes on demand as a response to a given stimulus pattern. System Theory as discussed in chapter 4 describes how an autopoietic system assembles a meaningful response by carrying out sequential selections (processing). A meaningful selection, we remember, is a selection that reflects the system's organizational setup (meaning). The rationale behind the selections, together with the strategy when to trigger them and when to

[34] Note that, as discussed in Chap. 3, today's enterprises are far from the ideal of an intelligent enterprise. For this reason performance issues do and will play a significant role in reality. The past indeed taught that collective performance improvements did lead to positive results. This is a fact that encouraged the emphasis on improving the collective performance of the individuals while the focus on coordinative efforts to improve the overall behavior of the company as a unity in its environment got lost. But it is exactly this 'System Behavior' that distinguishes intelligent from non-intelligent systems – it is thus an element we should expect to gain prominence rather in the future than in the past.

invoke follow-up steps, in the end controls the chaining of actions to a suitable response.

Borrowing the terminology from Computer Science (KOWALSKI, 1979 [42]), we call this the control of the process assembly:

The system's processing *control* is given by the rationale behind the selections together with the strategy when to trigger these selections and under which circumstances to invoke follow-up steps. [Def 5.2]

What is selected by the control mechanisms is actions, procedures or process fragments that have a known implementation – the Process Building Blocks (PBBs). In order to assemble 'enactable' solutions, the control mechanism has to constrain its selections to implementation level PBBs that are available as the system's microscopic procedural knowledge (see 5.3). We defined these PBBs as the constraints of the process assembly:

The system's processing *constraints* are given by all the building blocks – implementation level procedures – the system has incorporated (enactments available for disposition). [Def 5.3]

Note that this definition of constraints does imply that the more PBBs there are available – thus the more constraints there exist – the more freedom exists for any of the selections the control mechanism has to carry out. A language convention that may seem unfamiliar at first sight. If we however consider a system with no freedom at all, there is no need for the description of any constraints. For every new degree of freedom constraints are required that formulate or describe the use of the newly gained freedom. The more degrees of freedom there are, the more constraints have to exist to describe them – this corresponds to the understanding of PBBs acting as the system's constraints.

Given the introduction to constraints and control, it is obvious that it is their task to define the macroscopic behavior of an intelligent system. The interplay of control and constraints drives the assembly of a response process to a given challenge. They decide which microscopic procedures or building blocks are invoked and as a consequence last but not least define the system's overall behavior! Thus we define:

The system's knowledge from a macroscopic perspective is given by the set of constraints and control that enable the instantiation of those processes which coordinate and use the system-specific new degrees of freedom gained through system integration. [Def 5.4]

The control mechanisms select building blocks based on the assumption that the encapsulated procedures are 'enactable'. The system in a sense limits its scope in that it takes the integrity and complexity of the building blocks as given. Complexity in this way is very naturally handled without even trying to manage it. Encapsulated in functionally closed units it is broken down to simplicity: the PBBs may encapsulate arbitrary high- or low-level procedures that are assumed to be enactable where, at the time of assembly, their inner complexity – including the question of how the enactment takes place – is hidden.

The constraints or the building blocks are the elements that interact to propagate the stimuli – they define the space in which processes emerge. If we follow the discussion in chapters 4.1 and 4.2, the constraints are thus to be regarded as the system's structural components.

5.5 Declarative Processing

Classical Philosophy sees logic as the "... teaching of consistent and ordered thinking ..." (KUNZMANN, BURKARD, and WIEDMANN, 1991 [46], p. 13). Logic in this sense provides the techniques for the deduction of a meaningful response. Classically, logic was separated into the elementary teachings (discussing term, judgment and conclusion) and the teaching of methods (of investigation and proof). Based on this framework, the analytical philosopher and mathematician Gottlob Frege built a sophisticated symbolic logic. He formalized and standardized the elements of the classical framework and introduced such prominent concepts as the predicate calculus and quantifications (KUNZMANN, BURKARD, and WIEDMANN, 1991 [46], p. 219). According to Frege's idea, every logical expression, sentence or formula does not only have content but also meaning[35] – an understanding that gave rise to many years of mainstream artificial intelligence (AI) research that assumed meaning to be translatable into symbolic domains and knowledge to be representable in logic languages following so-called model theories. Based on these languages, world models can be formulated, stored and manipulated as well as conclusions, verifications or falsifications can be deducted formally.

[35] Frege assumed that logical expressions, sentences or formulas have (ger.:) "Sinn" and (ger.:) "Bedeutung". "Sinn" here has been translated into 'Meaning' in the sense meaning was introduced in Chap 4. "Bedeutung" here has been translated into 'Content' as it refers to a symbolic function it fulfills. Both 'Sinn' and 'Bedeutung' would in this context commonly be translated into 'Meaning'.

> *Providing this conceptual approach for the computer programmers, high-level, intelligent programming languages were proposed. The underpinning idea behind these languages was Kowalski's famous proposal to understand any computer algorithm to be composed out of logic and control: "algorithm = logic + control" (KOWALSKI, 1979 [42], p. 424).*

In his sense **logic** is the '**What to be Achieved**' or the **goal** to be computed whereas the **control** defines **how the computation actually takes place**. Kowalski suggests to separate logic and control requiring the programmer to declare the logic but not necessarily the control (which should be provided by the programming environment) – an approach widely known as "Declarative Programming". (TORGERSSON, 1996 [110])

Meanwhile the classic understanding of humans (or any other living system) processing information according to a formal logic or the brain being an information-processing unit has been widely dismissed by biologists (MATURANA and VARELA, 1987, [59], p.169). Meaning, as introduced in chapter 4, is engraved in the system's structural and organizational setup. It thus cannot be transferred even from one system to another; neither can it be transferred to a logic domain model. As a consequence, the same is true for system knowledge. Both, meaning and knowledge, have to be built-up self-referentially. Nevertheless does Kowalski's notion – given his understanding for logic and control – in a sense still stay valid for the processing, operationally closed autopoietic system: the incoming stimuli provide the system with a notion about 'What to Achieve' (which is slowly perceived through processing). It is the 'logic' driving any of the system's activities. Whereas the way 'how' the system responses to the stimuli pattern, how it carries out its sequential selections is defined by its own control (meaning). This is true even for the invocation of process building blocks as described in 5.4, just as it is also true for the way the system's participants (e.g. body cells) are invoked: by declaring **what to achieve** and **not how to achieve** it – **Declarative Processing**.

As already mentioned in 3.2 declarative processing provides a solution to the implementation of highest dynamics based on an almost static constitution. In addition it is a natural solution to handle complexity as described in 5.4. Macroscopic processing effectuated by macroscopic knowledge as discussed in 5.4 now is a solution to declarative processing.

Logic, control and constraints are the three factors in command of the system from a macroscopic perspective:

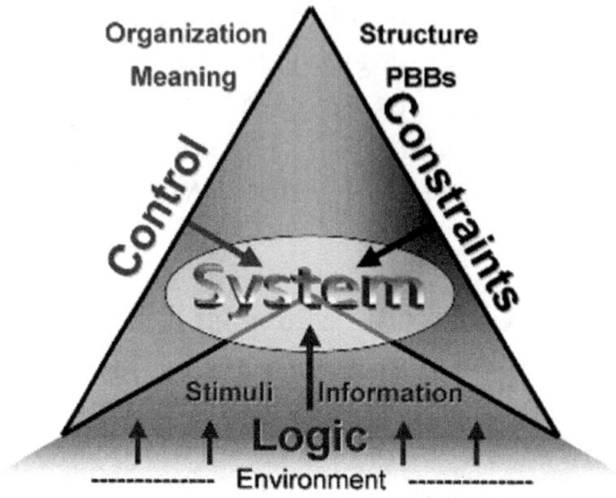

Fig. 5.1. Logic, control and constraints

5.6 Macroscopic – Microscopic Integration

The macroscopic perspective focuses on the system's behavior and it's macroscopic activities. With respect to them, knowledge forms the basis on which solutions are assembled and enacted. The microscopic perspective takes a rather mechanistic approach identifying the process itself as the underlying microscopic screen script defining behavior. Microscopically all macroscopic process building blocks (the constraints) are themselves physically mapped and executable microscopic processes. So is the control of declarative processing a set of microscopic (high-level) processes that finally carry out the selections of resources (PBBs) to form the macroscopic behavior.

The macroscopic assembly of building blocks finally results in a new implementation level process that is enacted to generate the macroscopic behavior. Leaving structural effects, parts of it may eventually be incorporated and reinforced in the intelligent system: these are process fragments with a known enactment or implementation level procedures – knowledge from a microscopic perspective representing new PBB's or new control procedures to be used in future for the macroscopic assembly of behavior.

Exactly this is Autopoiesis! During processing the PBBs are constantly reproduced. So is the system's assembly control that reflects its organizational setup. Knowledge is derived and managed self-referentially in an autopoietic 'System of Knowledge' constituted out of PBBs (the structural components) and the assembly control (the organization). It is a system constituted out of a network of processes that constantly reproduces its own structural components and its own organization.

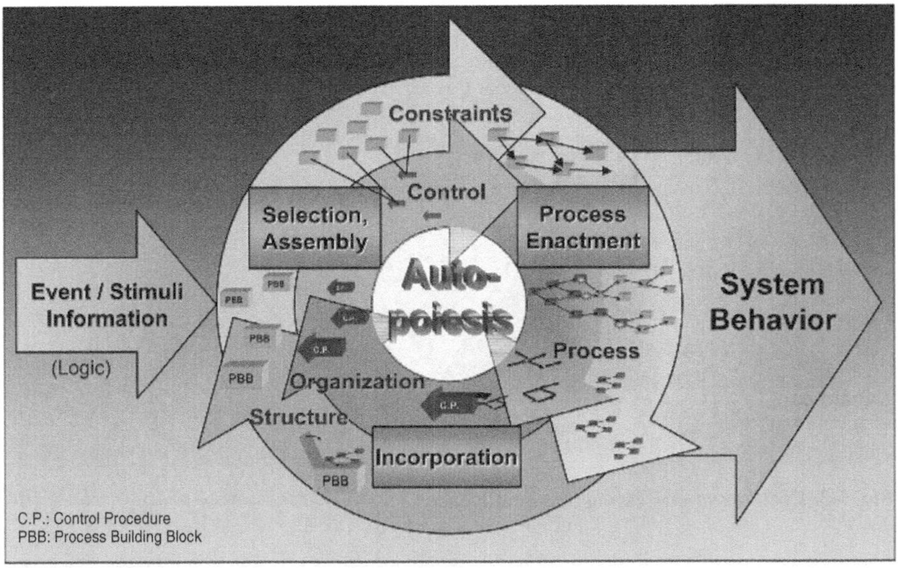

Fig. 5.2. Autopoiesis – the system of knowledge

5.7 Summary

System knowledge is an instrument and framework that defines the system's behavior. It has to be understood from a microscopic as well as from a macroscopic perspective. Microscopically, knowledge is given by implementation level procedures or enactable processes that coordinate all the system's microscopic actions. Macroscopically it is given by the constraints and control of the declarative assembly mechanism that provides a meaningful system response to a given stimulus (Declarative Processing). Knowledge is managed and derived self-referentially in an autopoietic manner, constantly reproducing, rearranging and redefining constraints (PBBs) and control.

Intelligence is a methodological framework arranging the physical incorporation of behavioral patterns (PBBs) in order to realize continuous adaptation. With the success of adaptation being perceivable and evaluable by the system it gives a clear direction to the ongoing development of PBBs and the control processes, without which precision could not be achieved and knowledge would not emerge.

Declarative processing is a natural solution to the handling of complexity and the realization of high dynamics. In addition intelligence, providing natural methodologies to select and extinct, ensures the necessary precision.

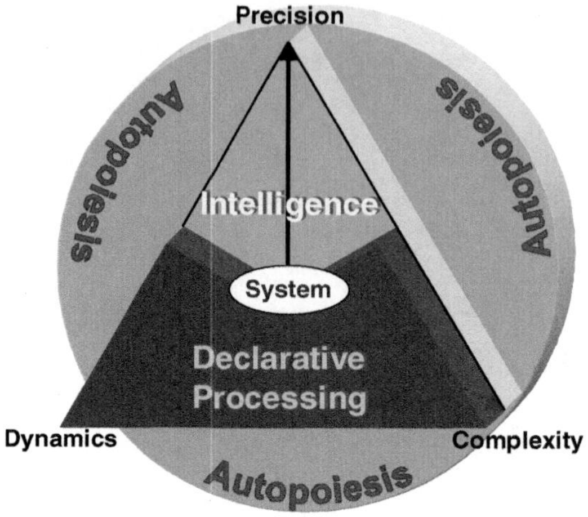

Fig. 5.3. Declarative processing and intelligence

It is worth mentioning that:

- Meaning cannot be transferred as such from one system to another just as it cannot be transferred to a logical domain model as it is physically engraved in the system's structure and organization.
- Knowledge, too, cannot be transferred as such from system to system, nor can it be represented by any language expressions, data collections and the like – it has to be built-up self-referentially.
- Neither can knowledge build-up without an intelligent framework that provides the necessary directional guidance nor do both of them exist without physical change mechanisms like Autopoiesis.

Chapter 6 The Intelligent Enterprise

The Application of the Derived Theoretical Framework to the Domain of Business and Engineering Management

The intelligent enterprise is characterized by the ability to optimize the balance of forces to which it is exposed on a high level of performance. While being flexible, adaptive and innovative, it is able to cope with high complexity, an enormous amount of information and a high variety of requested services even though it has to execute all its activities with great precision to achieve an outstanding overall execution effectiveness. This requires abilities to sense unbalances, perturbations and threats, react and adapt quickly, anticipate or even predict developments and last but not least actively influence or form the environment. The enterprise as a whole has to learn and refine its behavior within timescales much shorter than the employees are able to do so themselves. As participants of the system enterprise they are integrated in an entity that achieves more, adapts faster to changes and learns quickly even without the individual having the need to comprehend all economic challenges or learn and adapt that fast her- or himself. This provides security and a clear competitive advantage for the employee and thus last but not least is the only rationale for the existence of an enterprise in a liberal market of labor. As discussed in chapter 3, intelligence requires a sound organizational structure and physical setup. All implications derived for abstract systems in the previous chapters should now be transferred and discussed exclusively for the system enterprise.

6.1 Enterprise Behavior

Just like any other system the enterprise behaves by enacting processes. These include design, production, purchasing and logistic processes just as processes to evaluate employees, continuous improvement, accounting and others. Coordinating the company's microscopic activities these processes finally lead to an observable macroscopic behavior.

The system's participants are the employees, managers, CEOs, board members or in general anybody who is part of the enterprise's workforce. They are the enterprise's only sensors, enabling it to perceive perturbations and stimuli, as well as they are its only actuators to enact all microscopic activities. Macroscopic enterprise behavior emerges through the alignment and coordination of microscopic actions carried out by its workforce.

Following the discussion in chapter 3.2 typologically two distinct extremes can be identified: evolutional behavior on one side and intelligent behavior on the other side. Reality is located somewhere between these two extremes!

Companies showing evolutional behavior enact basically unchanged processes over long periods of time with environmental changes being taken into account only by changes in parametric values. Fundamental behavioral changes do not occur until the enacted processes show a significant lack of success and typically require disruptive events such as a change of management or even the suspension of operation. The system enterprise in this case is unable to implement behavioral changes without losing its character. From the perspective of System Theory, after such a disruptive event, the system enterprise even cannot be regarded as the same system anymore as its organization has been lost or replaced (s. Chap 4.1.4). Learning thus happens over generations rather than within the life cycle of the system: evolutional learning!

Intelligent enterprise behavior is characterized by the uniqueness of all enacted processes, each specifically tailored to the challenge faced. The enterprise behavior is constantly refined within the life cycle of the system: intelligent learning!

Intelligent behavior requires the employees to be understood and integrated as functionally autonomous, self-determined systems. Being themselves autopoietic systems, the employees thus naturally meet these requirements. They are not machines that have to be instructed every single step to take but in general are addressed in a declarative fashion: telling them merely what to achieve rather than how this is done[36]. It is this very human feature which is of most value to the modern enterprises – not their ability to aggregate or pool facts, their encyclopedic knowledge or incorporated expert knowledge – as only it allows the effective assembly of tailor made company processes.

---

36 Depending on the employee, the communication of task to be achieved may happen on a high or rather low level. In case a low-level task description is required the employee is told a full sequence of more simple tasks in order to achieve a larger goal. But still these simple tasks are at the end communicated in a declarative fashion as the employee in generally is not instructed exactly how his activities are to be enacted in detail – body movements, texts, facial expression or gestures for example are typically not instructed in detail but rather is it up to the employee to decide how they are to be carried out.

Not only does intelligent behavior however depend on capable participants but also requires those participants to be integrated in an appropriate manner. The employee's functional role at the generation and execution of processes as well as their integration within the organization as real knowledge workers has to be rewritten.

It is important to notice that enterprise behavior is phenomenologically distinct from the human behavior enacted by the company's employees. Enterprise behavior emerges through the coordination of numerous activities of employees in the enterprise processes just as human behavior emerges through the coordination of numerous activities of body cells in biological processes. Enterprise behavior is just as distinct from employee behavior as human behavior is distinct from the behavior of body cells. The emergence of enterprise behavior requires the coordinated participation of employees and thus mechanisms to establish processes as well as a suitable framework to guide and motivate the employees to participate.

6.2 Enterprise Intelligence

The intelligent enterprise is an organization that constantly refines the processes it enacts. It accepts the process to be primarily a coordination framework providing the underlying logic that couples the sequence of events and actions. A sequence that is – at first – independent from common needs, like repeatability, goals for effects to be achieved, consistency etc. Regarding each and every process as an one-time instance tailored to the specific needs of a given situation the intelligent enterprise achieves what is regarded to be the most fundamental element of intelligence: adaptation. Being successful in a world of rapid changes, uncountable new technologies, a high variety of customer requests, numerous distribution and supply channels and tough competition does require more than a set of streamlined putative optimized processes: it requires the intelligent generation of processes, allowing the enterprise to adapt to the changing economic needs and challenges it faces.

Enterprise-level learning brought to the point is the refinement of processes that coordinate the enactment of activities, adapting enterprise behavior to economic needs. Although effectiveness, profitability and consistency do not define the process in the first place, they finally have to define the direction of any developments. The degree of adaptation, and thus the success of learning, is to be measured by its effectiveness at minimizing irritations (s. Chap. 5.1). Losses, unhappy customers or lost market shares but also broken cooperations and unhappy employees are such irritations.

Once processes are enacted it is necessary that they are evaluated and that the underlying links and procedures chaining the building blocks are enforced or depreciated. The criteria of evaluation have to include such prominent economic benchmarks as the achieved profit per time – probably the most important economic indicator –, customer satisfaction or punctuality and degree of order fulfillment. Next to these main indicators a number of strategic parameters might be of relevance, such as the motivation of employees, the state and development of core competences or supplier relations.

As discussed in chapter 3.3, highly intelligent systems all distinguish themselves from less intelligent systems primarily by building on an advanced set of physical properties and abilities. The intelligent enterprise requires a sound physical and organizational setup. The physical framework an intelligent enterprise has at its disposition enables the ongoing refinement of processes and supports a dynamic enactment. It also has to implement the evaluation mechanisms, institutionalizing their rigorous application and assuring their results lead to consequences. More details on how such a physical framework will look like are presented in chapter 7.

The organizational setup of an intelligent enterprise has to regulate the integration and interplay of its employees. On the one hand the classic command structures can be regarded inappropriate for the integration of knowledge workers. They should contribute with their autonomous problem solving activities, independently seeking for solutions and taking whatever actions are required. On the other hand the danger exists that in a group of employees, who freely negotiate their coordination, enormous capacities are bound only by negotiation and coordination efforts. In addition, time and priority management, required to successfully schedule all steps throughout the process enactment as well as the management of resources, tend to be problematic. The intelligent enterprise has to relieve the knowledge workers of the burden to negotiate the coordination, providing them an environment in which they are exactly told what to achieve, in which time frame to achieve it and which resources they might use. This requires the enterprise to know about the capabilities of its employees as well as the time and resource demands of related tasks in order to arrange the process coordination and execution.

An essential point for the successful management of resources is the availability of estimations to the 'value added' achieved through the enactment of tasks and activities. Not only do they allow the enterprise to predefine the extent of resources that might be used by an employee enacting a task, but it also provides a possibility to estimate the contribution the employee was able to make by comparing the 'value added' achieved to the resources that had been used. With the need for unambiguous motivation schemes to keep the employees focused, rewards that are based on the employees' contributions are essential.

The establishment of flexible processes in intelligent enterprises demands for one more organizational consequence: a rigid project framework. In today's companies there exists a rigid framework of processes. In this framework projects are plugged in and carried out. Through the enactment of activities, which are coordinated by the company's process structure, the projects are moved from stage to stage until they are finally completed once they have reached the end of the process. But this rigid framework of processes provides more than only the coordination of the company's workforce! It is a stable structure in which the employees find a foothold and occupy their niche. It is a structure that gives them orientation and protects against swamping. Thus it is a highly important organizational entity that needs an adequate replacement in intelligent enterprises. Organizing the company's operations in a rigid project framework would provide many of the aspects the rigid process structure traditionally offers. With each and every activity being anchored in a well-defined project, these projects provide a system of rigid reference points to the employees. Processes can then be tailored for each and every project, dynamically and as one-time instances!

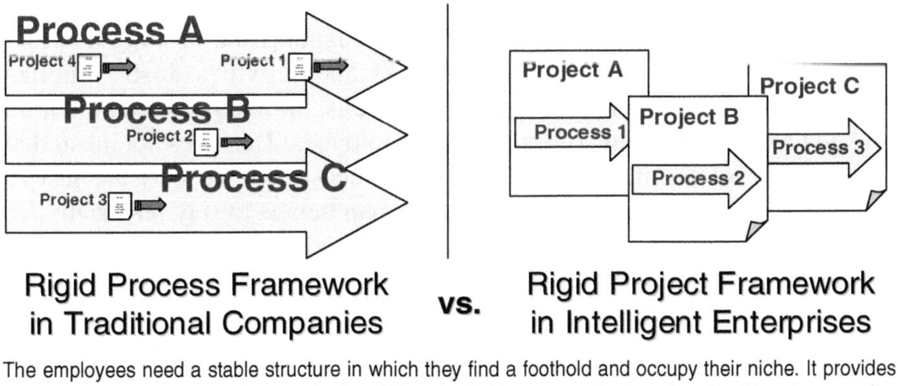

The employees need a stable structure in which they find a foothold and occupy their niche. It provides orientation and protects against swamping. What the rigid process structure is to traditional companies (Process-Driven Management) is a rigid project structure to intelligent enterprises (Project-Driven Management). Here, each and every stimulus is first assigned to a project; each and every activity is anchored in a specific project. Working in projects thus provides a stable environment and leaves room for flexible, tailor-made processes within the projects.

Fig. 6.1. Project- vs. process-driven management

'Enterprise Intelligence' is to be regarded as a methodological framework that institutionalizes adaptation through continuous refinement and evaluation. It allows the company to cope with unfamiliar market situations, adapt to new approaches, strategic and tactical concepts and develop core competences. It is the "Innovation Engine" that enables a constant repositioning of the enterprise and thus provides the necessary competitiveness in a dynamic market environment. Intelligence is necessary to develop the enterprise's cognitive domain that enables it to respond consciously to numerous economic stimuli.

6.3 Enterprise Knowledge

As introduced earlier for abstract systems, knowledge complements intelligence. Successful enterprise behavior only emerges if both components, enterprise intelligence and enterprise knowledge, exist. Where intelligence provides the methodological, physical and organizational framework, knowledge is the set of procedures, practices and solutions empirically built up to fill the framework. A company's knowledge is defined by its history, the niche it occupies, its customers, competitors and suppliers, its employees, the technology it applies and other elements that define its context. Enterprise knowledge thus reflects the very specific character of the enterprise and drives its behavior in any specific situation.

Enterprise knowledge thereby is phenomenologically distinct from human knowledge or in particular the employees' knowledge. The enterprise learns – builds up enterprise knowledge – through adapting and refining its processes that coordinate the interplay of tasks and activities. Just as human beings do not learn by improving their body cells, the enterprise cannot build up knowledge by teaching or training its employees. This does not mean that efforts to develop employees' skills are unnecessary – no they are just as necessary to the company as it is to the human beings to stay physically fit, condition the body's physical cell structures and follow a healthy diet. A well-conditioned, highly skilled and effective workforce is the basis for physical performance but not for intelligent behavior. Most of the traditional Knowledge Management efforts metaphorically speaking take the enterprise to the gym to physically train and shape – which definitely improves it's performance to a certain extent – but they don't take it to the classroom to sharpen it's intellect.

Following the discussion in chapter 5, enterprise knowledge will basically be formed by microscopic constraints as well as the macroscopic control.

The interplay of constraints – *the building blocks out of which processes are assembled* – and control – *the assembly mechanisms chaining building blocks to a process* – finally results in highly specific processes tailoring behavior to the specific needs of a given situation.

a) Constraints

The building blocks, to which the control mechanisms have to constrain their selections, are procedural entities that the system enterprise has available for disposition. In the simplest cases these are the capabilities of staff members, which they put at the enterprise's disposal. A capability encapsulates any desired set of actions that is enacted by the employee providing it. The enactment of these encapsulated actions is not under the enterprise's control – they are out of its scope. In case the enterprise asks for the enactment of an employee's capability, the employee will decide her- or himself what is to be done and especially how it needs to be done. Although the employee – or any of her/his colleagues – might be able to further decompose her or his capability, offering an even more fundamental set of building blocks, the system enterprise itself, is unable to further decompose a capability. For the enterprise a capability thus is a generic element with atomic character: an **atomic process building block** (atomic PBB). From a system theoretic perspective, the capabilities are even more than just building blocks. They define the interface between system 'employee' and system 'enterprise', enabling the communication of what to achieve. They are the interfaces that mediate between enterprise and employees.

Next to atomic process building blocks a company may have internalized procedures that integrate several generic building blocks. These larger fragments of a process typically have been identified as best practices or feasible standard solutions for a given challenge. Used as generic units this greatly reduces the assembly effort and allows the formation of complex processes[37]. Although they integrate several capabilities from possibly multiple employees involving an arbitrary number of actions they still may be seen as

[37] The assembly of complex processes is technically always limited by time constraints. This is not only the case on enterprise level but is also true for human beings. Somebody who has strongly internalized mathematical methods and approaches to common problems (someone who has pre-constituted compound PBBs available for disposition) will e.g. be able to address even complex mathematical problems, whereas somebody who has not internalized these methods and approaches will not easily tackle even simple problems – no matter how intelligent she/he might be. The time for inference, deduction or assembly of complex processes obviously has to be constrained in order to keep the system responsive! Thus it is often the availability of compound PBBs that 'constrains' the ability to respond in situations requiring complex behavioral processes.

a single generic building block by the enterprise, as long as this set of activities

- is preconstituted and ready for enactment
- enables the integration into a larger process.

Unlike atomic process building blocks these **compound process building blocks** (compound PBBs) emerge as products of the system's own activities and thus may theoretically also be disintegrated by the system into its components.

Apart from these two types of building blocks, it is hard to imagine any other element that would qualify as process building block. This is a natural consequence of the fact that the system's processes are triggered and enacted only by its actuators and sensors. These are the employees, managers, CEOs, board members or generally speaking the enterprise's participants. Hence will every process building block be directly derived out of one or several capabilities provided by the participants[38].

Examples:

- An e-business interface that captures the customer's orders and triggers follow-up steps in the order handling process is a capability provided by the employee who programmed or provides the interface.
- A set of automated production sequences is a capability provided by the worker who runs the machines or possibly – in case of full automation – by the automation engineer.
- The delivery of semi-finished products from one production site to another by an external forwarder is the capability of the employee who contacted and ordered the forwarder.

As discussed in chapter 5, the process building blocks are to be regarded as the structural components of a system. Consequently the capabilities are the structural elements of the intelligent enterprise. More than regarding an enterprise as an economic or social entity, it should be regarded as a set of capabilities that are paired up to achieve economic success for its participants, the employees.

[38] Those external partners might be regarded as a special case, who highly integrate with the company's processes in the sense that they provide capabilities that are directly addressed by the company's control mechanisms or that directly address the control themselves. They rather play the role of temporary participants than that of external partners.

b) Control

According to definition 5.2, the control procedures determine the chaining of building blocks to a complete process. The enterprise's control thus defines how an employee's capabilities are integrated and involved in a business process. In order to ensure the quality of the assembled processes the control has to make use of and learn from the evaluation mechanisms provided by the intelligent system. It has to utilize known best practices and avoid depreciated solutions. By invoking the capabilities it plays a central role for the declarative processing paradigm: it defines the propagation of stimuli from capability to capability. Although the control – by addressing the employees' capabilities – will communicate only a notion of 'what to achieve' not 'how' this is done, it is still concerned with 'how' capabilities are to be chained. Thus the control itself is a 'solution provider' to the challenge 'chain a process' and can itself be addressed in a declarative manner. As a consequence the control procedures can be regarded as one or several designated process building blocks.

In today's companies, the middle management can be identified to design and establish the existing process landscape. Indeed, today it is the shop floor managers, department heads, division managers, directors of operations, executives and so on who provide the capabilities – the control procedures or designated control process building blocks – to establish their company's processes. These processes, however, are established once and used unchanged for rather long periods of time. Once established, the necessary control operations are limited to process monitoring and eventually to routing decisions that trigger predefined process branches depending on a set of well-defined parameters. The processes are hardwired! A real-time control that chains employees' activities to executable enterprise processes on-the-fly hardly exists.

This traditional approach does not build on natural self-organization but requires the involvement of managers to fully comprehend not only strategic issues but also the full implementation level requirements, at all involved tasks as well as their failure and success rates under a variety of circumstances. Preparing a set of hardwired processes that lead to successful enterprise behavior not only at the time of preparation but even in the near future requires in addition that dynamic environmental changes are predicted correctly by the middle management. Consequently, today's enterprises do not only strategically but also tactically fully rely on a group of employees. The enterprise learns – adapts its behavior – as quickly or slowly and as efficient and effective as they learn and adapt. This strongly contrasts to the expectation we have from highly developed intelligent systems, which develop, adapt and learn by scales faster than their participants. In chapter 3.2 this was discussed as an evolutional necessity, providing higher-level systems with the necessary competitive advantage against lower-level systems acting as sole players.

Intelligent enterprises thus have to develop a control that not only facilitates but even enforces the self-organization of capabilities. It has to enable the auto-assembly of employees' activities to a process that is tailored to the given challenge, the specific circumstances and the internal situation in the enterprise. But it also has to support the self-production of capabilities, which are the company's structural elements. This includes

- the formation of compound PBBs,
- supporting the communication and instantiation of new atomic PBBs in interactions with the employees, and
- the instant integration of new PBBs.

How such control mechanisms are physically implemented and realized should be discussed in chapter 7.

6.4 Declarative Processing on Enterprise Level

Declarative processing emerges in an intelligent enterprise as soon as constraints and control are put to work. It provides a suitable response to a given set of stimuli. In the business domain, this trigger has to be interpretable by the enterprise as an achievable objective, like "deliver products", "arrange repair/service" or "order supply". For simplicity, we call such a well-configured set of stimuli a '**service request**' (SR) whereas any other stimuli not implying a perceivable objective are called '**challenges**'.

With every enterprise being a functionally closed system (s. Chap. 4.3) bound to the physical laws that govern such systems, it can only yield profits and produce assets if it turns over a substantial amount of products or services (see "The Engine Enterprise", Fig. 1.2)[39]. Thus it is imperative for the objective that is given in a service request to be of a specific nature. "Making profit" cannot be an objective at first-place – rather is the profitability of the assembled and enacted processes a fundamental need for the enterprise's

[39] Being a functionally closed system, an enterprise, earning stable profits without turning over products or services, cannot exist, as it would violate the 1st Law of Thermodynamics. Furthermore, the efficiency of an enterprise in earning assets is limited to Carnot's law (2nd Law of Thermodynamics).

existence and thus one of the major parameters to be continuously evaluated and monitored. The profitability of previously enacted processes will influence the selections that lead to the assembly of a new process in future. Yielding profits is the fundamental goal of declarative processing as a whole. If a given service request requires a non-profitable process to satisfy the service request's objective, it would in consequence be highly likely that the control mechanisms are unable to assemble a process at all – such a service request would simply be out of the enterprise's cognitive domain!

Declarative processing as described in chapter 3.2 and later chapter 5.5 would, once the service request is perceived, result in an immediate invocation of all required employees, just at the time of need. "To Do's" would be communicated directly, without mediation through the middle management. Ideally activities would start right at the time of need, coordinated and in parallel, leading to an immediate response to the service requests. The middle management would not be bound in endless meetings, resolving coordination issues, or tactical planning; it could rather work on strategic issues that drive the enterprise. However, the increased complexity of circumstances in the higher level system 'enterprise' – compared to biological systems – does require to have some profound issues resolved first:

- The interactions to communicate the objectives (the "What to achieve") are complex and need substantial descriptions regarding the context of the situation. This requires advanced communication techniques facilitating the interpretability of content for all participants and enabling the control procedures to enact content-based selections.
- A direct invocation of employees binds significant resources and thus causes extensive financial losses if not enacted to the satisfaction of all parties. The safe invocation of employees requires a careful planning including the management of resources as well as task scheduling capabilities to avoid time and capacity conflicts. These are two additional functionalities the control mechanism or the capabilities have to provide to facilitate declarative processing.

Declarative processing plays a central role in all intelligent systems. It provides a solution that realizes dynamics, allows the handling of complexity and at the same time is bound to the results of the system's evaluation mechanisms guaranteeing the necessary precision in all assembled processes. Technical approaches to declarative processing will be discussed in chapter 7.

6.5 General Implications and Organizational Consequences

This new conceptual approach to analyze and understand the intelligent enterprise leads to a number of implications and consequences. Three of the most important should be discussed in this chapter.

6.5.1 The Role of the Employees

The special role the employees play in any system theoretic contemplation requires some further discussion. Not at least, because even Luhmann's postulation, classifying the human beings involved in the formation of a social system only as the system's environment, was heavily criticized (MINGERS, 1994 [62]; WHITAKER, 1995c [118]), although he provided a comprehensive theoretic framework to support his argumentation.

And indeed, the employee is of course more than just the enterprise's environment! Any interaction, in which the enterprise enters, involves its employees. Any process that is triggered is triggered via the employees. Any action is carried out by addressing the employees. They are the enterprise's only actuators and sensors. If regarded as their structural elements or not, they define its interaction level: questions, offers, requests and claims are directed to the human beings, who represent the enterprise and donate its visual appearance.

But the employees are more than this! The employees provide the atomic PBBs and are thus the enterprise's only source generating 'value added'. Being bound to the imperative of profit-making (s. Chap. 1) the generation of 'value added' is an absolute necessity for every company. But what is 'value added', how is it measured and described? It is beyond the scope of this work to discuss 'value added' scientifically. Thus any introductions and descriptions given here should be limited to the bare minimum.

By tradition, 'value added' achieved by a business operation from the perspective of business administration is the contribution this operation made to the national income (GABLER, 1995 [30], p. 3757). It is given by the total proceeds minus goods and materials employed[40] and thus is closely related to the operation's profit, which is 'value added' minus labor, capital and operational costs. Broken down to a single product, 'value added' conse-

[40] Defining 'value added' in terms of total proceeds minus goods and materials employed is typically limited to the producing Industry. For simplicity this definition here is extended even to the service industry where goods and materials can typically be ignored. This definition would be problematic for all cases in which 'value added' is brought in relation to the input factors, which is not the case here.

quently is the product price the customer is willing to pay minus all goods and material costs involved to produce the product – or in other words, it is the price the customer is willing to pay for all operational steps required to sell the product. Last but not least, 'value added' thus is generated through these operational steps or the involved tasks and activities. This gives rise to a second, rather microscopic perspective: 'Value added' is gained through activities that form, develop or prepare a product in all its features and its environment (sales channels, marketing, etc.) in a way that is valuable for the customer. An activity generates 'value added' if and only if its effects lead to the customer's willingness to pay for the activity! The 'value added' achieved through an activity thus is exactly the price the customer is willing to pay for its results. The activity leads to profit if and only if the sum of all costs involved at its enactment is smaller than the 'value added' it achieves. Being forced to run profitable an enterprise can only realize 'value added' if the required activities do not cause losses – any other 'value added' is hypothetical!

This naturally leads to a need for figures that quantify 'value added', enabling the enterprise to estimate profit per activity and allowing it to identify realizable 'value added' and screen out non-realizable 'value added'. As the customer her-/himself lives, works, earns and spends money in an economy her or his willingness to pay for product features will be driven by reference points and indicators that she or he perceives from the market. As discussed in chapter 1 the demand curve – which graphically illustrates the number of customers that are willing to buy for a given price – significantly tilts with increased competition leading to the fact that most customers are not willing to pay much more than absolutely necessary to produce a product. Translated into the world of activities, this indicates that 'value added', which may be achieved by an activity, varies only in a tight interval around the typical cost benchmark defining the industry standard of the given activity. This indicates two major consequences for the enterprise and its employees:

a) As there are usually many ways to generate a product feature or a functional requirement in the customer domain, great care has to be taken to select the right! A sub-optimal product design or production technologies will not be successful even if each activity carried out meets or beats the industry standards. Needless to say that this requires processes to be rebuilt permanently to the availability of new approaches, new technologies and new capabilities.

b) Once a capability is communicated and used for a range of purposes the employee has to ensure that her or his task enactment meets or beats the typical learning curve for the involved task. Otherwise the capability's results soon can't be sold anymore and any 'value added' is hypothetical. How learning curves can be used to estimate the 'value added' and profitability of a task is described in chapter 7.

Approaching enterprise processes from a microscopic perspective or analyzing them in terms of capabilities and their underlying tasks, provides a great conceptual framework to think of and implement any kind of automation. Automation, as a matter of fact, is one of the most important and most successful paradigms at work in all developed countries. But automation as such does not generate 'value added'. Rather does the market price of a product tumble severely with the degree of automation: on the one hand, the 'value added' generated through an automated production of products decreases significantly with the degree of automation as the competitive market soon educates its consumers to rethink their value systems, adapting their willingness to pay only for the decreased production costs. On the other hand, competition does force every enterprise to thrive for competitive advantages leading to increased automation throughout all industries' business and production processes. Not automating an automatable task will soon prevent a company from being able to realize 'value added' at all. Thus every enterprise needs the strong commitment of all employees to automation – not to generate value but to stay in business and think of new possibilities to realize 'value added' and profit.

'Value added' is not achieved easily or automatically. Rather does it require human creativity, ideas, intellect and – possibly more than anything else – human efforts to realize 'value added'. Elements that can only be provided by the employees.

All this indicates the great importance of employees and the special role they play – without employees no company! Employees and enterprise are not really just interpenetrating systems[41], as their relationship is rather unbalanced. The human beings playing the rôle of employees may very well exist without an enterprise but not vice versa[42]. Nevertheless, it has to be understood that the employees are not a physical part of the system enterprise –

[41] Interpenetration is a term originally coined by Talcott Parsons. Later Luhman adopted this term to describe the reciprocal dependency of humans and social systems. Both of them exist only due to the other even though both are only environment to the other. (s. REESE-SCHAEFER, 1992 [81])

[42] This should not imply that the enterprise strongly depends on each and every employee! The enterprise of course can compensate the loss of single employees and the capabilities they brought in.

they are themselves autonomous entities, functionally closed autopoietic systems that can only be motivated to take part and act on behalf of the system but they can never be forced to do so. The employees themselves obviously face a challenging life and participate in an enterprise not without purpose. They expect security, protection and an environment in which they are able to contribute and achieve reliable incomes – competitive advantages along their way through life.

6.5.2 Evaluating Enterprise Intelligence

Any managerial efforts depend on well-defined criteria and parameters that allow a presentation of evidence, quantifying the success or failure of implemented measures. Thus it is not surprising that investors and managers but also partners and stakeholders demand for trustable indicators and parameters that measure the company's intelligence and its knowledge assets with sufficient precision. As mentioned in chapter 2 these demands reach even further, asking for a comprehensive balancing of intangible assets and intellectual capital. Obviously, the fundamental phenomenological differences between the production factors in these cases have widely been ignored. Unlike the production factor capital, which in the form of financial assets, cash or products is the operating medium the engine 'enterprise' uses, all the other production factors do not flow in looped circuits. They don't follow the laws of flow conversation and their entities cannot be expressed by simple numerical parameters. Thus there is no way to balance them in a conventional manner. Land or labor for example only appear in the enterprise's balance sheets in terms of how much capital they bind or drain but not how suitable they are or how strongly they support the given set of production or business processes. Exactly the same is true for knowledge and intelligence! The best to be achieved thus is a set of standardized parameters that allow an objective, industry-wide benchmarking. It is then up to financial analysts to calibrate these parameters to the potential business success or profit in order to obtain hypothetical financial values which investors might be willing to pay on top of a company's book value.

Even with this reduced demand, evaluating and benchmarking enterprise intelligence or enterprise knowledge however will be at least as difficult as evaluating and benchmarking human intelligence or human knowledge. Already the short discussion in chapter 5.1 indicates, which fundamental difficulties researchers have to expect when they explore this domain.

It is out of the scope of this work to discuss potential parameters and how they are gained. However, at least some of the more and some of the less relevant areas should be mentioned out of which a parametric analysis could be achieved:

Most important factors

- Quantifications on the size of the cognitive domain that the enterprise is able to access with its knowledge (constraints and control). This is probably the single most important indicator of the company's overall adaptability to common challenges.
- The degree to which the system can cope with unfamiliar and untypical tasks (ability to extend its core competences) and the average temporal development of achieved profit per process from the moment the service request is new and unfamiliar up to the moment it is well-known to the enterprise (learning curve).
- The average 'value added' and profit achieved over different areas or segments of the enterprise's cognitive domain (distribution and depth of the enterprise knowledge).
- The typical temporal development of the degree of automation in assembled processes as well as the availability and use of compound process building blocks over different areas or segments of the enterprise's cognitive domain (effectiveness and ability to internalize).
- The average time required to develop a cognitive domain in new strategic segments catering challenges that have previously been inaccessible for the enterprise (ability to develop new competences)

Points that are less important for the evaluation of enterprise intelligence

- The employees education level, IQ, skills and the like which are phenomenologically distinct from those of the enterprise.
- IT infrastructure or the application of software packages like information management systems, data warehouses, ERP systems, product information systems, and so on, which by tradition show a bad correlation to enterprise success.

Both are more concerned with the company's physical condition, which – without doubt – does have influence on its overall performance. But they measure more the company's fitness level or health situation rather than its intelligence.

6.5.3 Intelligent Interactions, Knowledge Types and Rational Behavior

By tradition, intelligence was believed to manifest in rational interactions, which all share a highly prominent link to the underlying behavioral models:

explicit knowledge. However, are intelligent interactions always of a rational type involving explicit knowledge or does another form of intelligent interaction exist? What is explicit knowledge and how does it emerge on enterprise level?

a) Intelligent Interactions Among Companies Forming Higher-Level Systems

In many cases companies interact with each other as they have committed themselves to form long-term cooperations in order to integrate their often complimentary abilities or merge their volumes to achieve scale effects. Typical examples are technological or financial cooperations or cooperations that lead to a strong integration of the supply chain. In all these cases the companies interact on the basis of a long-term and symbiotic collaboration virtually forming a distinct, higher-level system. As the discussions in chapter 3,4 and 5 have not been limited to a specific system level, the same rules apply to these higher-level systems: the system's intelligence is characterized by its ability to adapt its processes to the specific challenges given. This requires control mechanisms as well as enactable capabilities, which encapsulate any desired set of activities that can be chained to a larger process. The capabilities have to be brought in by the interacting companies playing the role of the system's participants. Although the interactions may be even more dynamic, the capabilities more complex and the control might have to cater for higher demands there is no fundamental difference to the discussion on intelligent enterprises. All interactions in this sense are just as intelligent as the interactions humans carry out in the system enterprise – with or without the involvement of explicit knowledge.

b) Knowledge Types and Rational Behavior

As described in chapter 2.4 it has been common practice to typify knowledge into distinct classes or mutually exclusive property pairs in efforts that should provide well structured and analytical approaches facilitating an easier access to the phenomenon knowledge. Most of these classifications, as already Probst and Romhardt pointed out, are however of little practical relevance (PROBST and ROMHARDT, 1997 [78]; ROMHARDT, 1996 [84]). Any discussion on rationality however indicates that one of the most prominent typifications is to be analyzed first: the classification of knowledge into an explicit or implicit type.

Explicit knowledge is commonly referred to knowledge of a – as Ikujiro Nonaka puts it – "formal and systematic" type (NONOKA, 1991 [65]) that can be articulated and communicated in the form of, e.g. oral or written texts, formulas, specifications, descriptions, diagrams, drawings and the like. It is the basis for rational behavior.

In order to be characterized as explicit, the knowledge as well as the intelligent system in which it exists thus have to meet several necessary criteria:

I. The existence of a content language
Articulation requires the existence of a high-level content language which enables formal descriptions that can be perceived and verified[43] by other intelligent systems. This requirement alone eliminates most biological systems to qualify for the development of explicit knowledge, as only one species was able to develop high-level content languages[44]: human beings.

II. Logical content and truth-values
The attributes 'formal' and 'systematic' require a type of knowledge that is bound to logical content and logical truth-values. In all biological role models however knowledge basically occurs as a procedural type driving behavior. As discussed in chapter 4.6.3 logical content or truth-values cannot even be memorized as such but require strongly enforced processes that lead to repeatable response pattern given similar stimuli pattern. A child, for example, may incorporate "1+1=2" in that it has learned that a behavioral response which can be interpreted as "2" is highly successful – causes minimal irritations – in situations where the stimuli can be interpreted as "1+1 equals?". The primary set of logic relations and truth-values is thus highly empiric. More than this, these high-level response patterns leading to the appearance of incorporated content or truth-values are only an extremely small fraction of all incorporated behavioral patterns – rather an exception than the rule.

III. Conscious awareness of mental states
The articulation of logical content, truth-values and relations furthermore requires a conscious awareness of those high-level behavioral patterns that represent them. A parrot that has learned to answer with "2" to the question "what is 1+1?" does not articulate its explicit knowledge but shows a rather simple response behavior to a given stimulus. Articulation however is a highly developed response behavior that requires the system to be able to reflect on and predict its own expected behavior – which reflects its incorporated knowledge – and the ability to translate the results, upon which the

[43] Verification of course does pose a further requirement to the content language! The language has to provide the necessary framework facilitating logical deduction.
[44] Languages that provide the necessary framework for logical deduction.

system has gained awareness of, into formal representations that can be communicated.

Implicit or tacit knowledge are all those incorporated behavioral patterns that cannot be classified as explicit, or in other words knowledge that cannot or at least has not been articulated.

Explicit knowledge thus is a very special type of knowledge. As already Polanyi pointed out, "while tacit knowledge can be processed by itself, explicit knowledge must rely on being tacitly understood and applied" (POLANYI in BOTEZ, 1998 [13]). Thus the attribute 'explicit' rather refers to an additional feature, which implicit knowledge has to have in order to be called explicit. "A wholly explicit knowledge is unthinkable" (POLANYI in BOTEZ, 1998 [13]). Regarding the list of criteria the attribute 'explicit' has to fulfill, it is obvious that all of them may exist in intelligent systems in a more or less developed form. This implies that there is no sharp demarcation-line between explicit and implicit but rather a scale of explicitness that provides many shades of gray.

In the discussions of chapter 3 to 5 no consideration has been paid to the distinction of knowledge types or to the criteria that characterizes explicit or implicit knowledge. The primary focus thus has been the more general implicit knowledge.

Explicit enterprise knowledge requires a conscious awareness of the system enterprise in that it is able to reflect and predict its own behavior. This requires constraints and control, which enable the successful instantiation of processes that simulate and predict the system's own assembly patterns. Furthermore, it would require constraints and control, which allow the assembly of processes that then would abstract these predicted results and translate them to a formal language. Nor a conscious awareness, neither abstraction mechanisms and a formal content language do exist today and cannot be expected to exist in the near future. Explicit knowledge does require highly developed intelligent systems and their highly developed properties – it is thus unreachable for enterprises that have just started to develop their intelligence. Hence, if we talk about Knowledge Management on enterprise level, we talk about managing implicit knowledge. Explicit knowledge on enterprise level is still beyond our imagination.

Chapter 7 Realizing Enterprise Intelligence

A Deeper Discussion on Concepts and Consequences to Realize the Intelligent Enterprise

Modern enterprises have to overcome numerous challenges on their way to realize and develop organizational intelligence. Among them are deep managerial changes away from the traditional rigid process frameworks that need to be replaced by project frameworks, the abundance of traditional command structures or the installation of a new understanding of the employee and her/his role. But none of the challenges faced is as demanding and difficult as the implementation of declarative processing, the self-generation and self-organization of building blocks to a process, enabling intelligent and dynamic processing.

In order to master this challenge the system enterprise – just like any other intelligent system – needs a suitable setup providing it with the advanced physical features and properties that characterize highly developed systems (s. Chap 3.3, Chap 6.2 and Chap 6.3). Probably the most important element to define the system's physical setup is the IT system in use. Despite all the criticism articulated by the organizational learning community (s. Chap 2.5), IT systems play a fundamental role when it comes to realize intelligence on enterprise level: they offer the means to channel information and communications, capturing and propagating stimuli to link sensors and actuators. IT systems thus will have to form the necessary core facilities, playing the role of the company's central nervous system.

In the course of the research carried out, an IT prototype was developed to show the feasibility of putting the proposed theoretic framework into practice. The prototype, called GCEN (Globalized Company Engineering Network), was planned to target the globally distributed engineering and design domain of an enterprise, for which the strong need to tailor-make processes is obvious. This chapter will discuss the developed prototype and its functional principles.

7.1 Objectives and General Implications

The overall objective to be achieved by the GCEN prototype was a technical implementation to support the intelligent coordination of employees' activities in dynamic, declarative processes for the engineering and design domain. This coordination requires communications to be triggered and managed as they last but not least carry the process.

Communications,

- in which employees (as the system's sensors) post service requests and challenges to the system, which they have become aware of;
- in the form of "To Do's" posted to the employees (as the system's actuators);
- in which employees state their capabilities (PBBs).

A particular process is given by a particular set of communications. The 'constraints' and 'control' that enable the dynamic and goal driven assembly of such a set of communications are the focus of the GCEN prototype.

Given the discussions in the previous chapters, this requires:

- a core facility in which the communications manifest and propagate;
- generic, task based process building blocks (PBBs) and a suitable framework in which they are developed;
- a declarative processing environment, which carries out the selections that assemble PBBs to processes;
- task scheduling[45] and resource management abilities;
- a framework in which continued evaluations are carried out and the system's organization as well as structure are continuously refined (the Autopoietic Framework).

[45] Scheduling here is to be understood in the sense of planning and arranging the application of given facilities and resources to avoid conflicts during the enactment of enterprise processes resulting from existing time or capacity constraints. Checking the availability of facilities or resources and reserving timeslots in an environment where multiple processes are chained in parallel is a non-trivial problem (dead-locks), so is the optimization of the time slot arrangement – both are challenging research topics in Operations Research and Computer Science. Note that this understanding should not be confused with the traditional understanding of 'Scheduling' in Industrial Engineering, where it is regarded as the coordination of supplies and production plans as well as the correct timing and distribution of products to the sales channels in order to fulfill existing and planned orders.

7.2 Communicating Capabilities

The GCEN prototype has to support the communication of capabilities. They are the building blocks or structural elements of the system and thus need to be captured and physically implemented. With the capabilities, the employee specifies what she/he is able and willing to do (task), under which circumstances to do it, what is required to do it and which results her/his activities will produce. Thus each employee will provide a number of PBBs each reflecting the different tasks or abilities she/he offers to the enterprise.

Being a structural element, the PBB has to interact with its environment, allowing it to propagate the chain of events and actions and playing an active role within a process (s. Chap 4.1.2 and Chap 4.2). As described later (s. section 7.3) it has to support the assembly process actively, evaluate the situation in which it has been invoked and place follow-up events. During the enactment the PBB needs to communicate to the employee asking for the enactment of specified tasks – tasks that the employee himself previously defined with the implementation of the PBB.

For practical reasons the PBBs thus have been implemented as autonomous software agents, which are invoked on demand, carry out any desired reasoning, enact simple actions (database manipulations, sending e-mails, posting of "To Do's" and the like) and post follow-up events.

7.2.1 The Service Requests, Fact Sets and the Ontology

The PBB in general is invoked by handing over a specific 'Service Request' (SR). The SR in a sense reflects the stimuli upon which the agent acts. In the course of the agents reasoning the SR is interpreted and evaluated. The SR communicates a picture of the given situation and its context. Technically it is realized by an arbitrary number of sensor names and sensor readings, which is communicated as a list of 'fact names' and 'fact values'. The 'fact names' are simple string expressions whereas the 'fact values' may be arbitrary objects. For technical reasons the object type of the sensor readings – or the 'fact value' – is included in the service request, although the system is not strongly typed. A typical set of facts may look like this:

Sensors: Fact Names		Readings: Fact Values
objective	-	deliver
product	-	UHT Milk
quantity	-	1000
customer	-	Superstore Lt.
due date	-	1.8.2005

Fig. 7.1. A typical set of facts in a 'Service Request'

The facts in the service requests are grouped in sets, the so-called 'fact sets'. This allows to communicate multiple objectives including their contexts or the integration of the process context into the service requests ("What happened in the PBBs before or after the one receiving the service requests?", "What were their objectives and results?").

The existence and use of particular sensors – or in other words: the 'fact names' – as well as their readings – or the 'fact values' – is published and documented in a simple ontology. The ontological expressions are structured in tree sets where each father node defines a term that can be used as a 'fact name', its child or in general lower-level nodes define the 'fact values', which are typically considered.

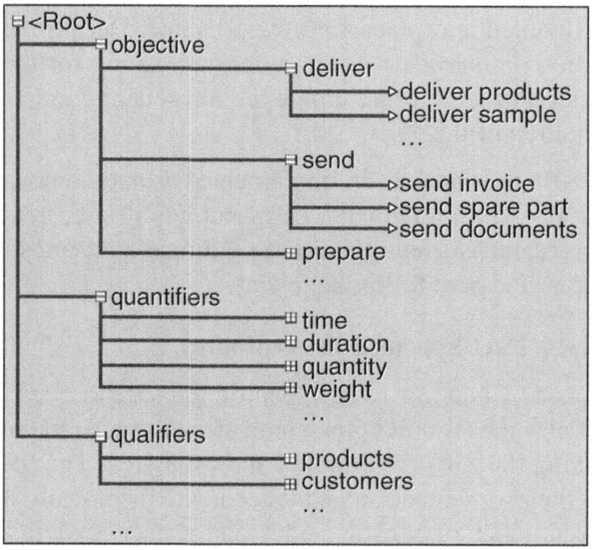

Fig. 7.2. The ontology

With each term or expression the ontology in addition captures its object type, a simple thesaurus and eventually a list of links to other terms that typically are required in its context. GCEN does functionally not depend on the ontology in the first place – it can handle an arbitrary set of fact names or fact values without having them registered in the ontology – however, it does help to standardize the terminology used by the employees. It also supports the users at formulating a SR. In addition the ontology of course does open the door for some advanced techniques such as thesaurus based matchmaking or term translation.

The SR is a well conditioned and machine interpretable set of information used to communicate the objective and context of a given situation. It is the exchange format used to handover information from PBB to PBB.

7.2.2 The PBB – Functional Description and Reasoning Engine

Each capability is identified by a name, carries a short natural language description and pursues a specific goal (a 'fact value' located in the ontology under "objectives", like e.g. "prepare shipment papers", "order forwarder" or "check availability"). It is these and some additional technical information the employees have to provide first, when communicating their capabilities.

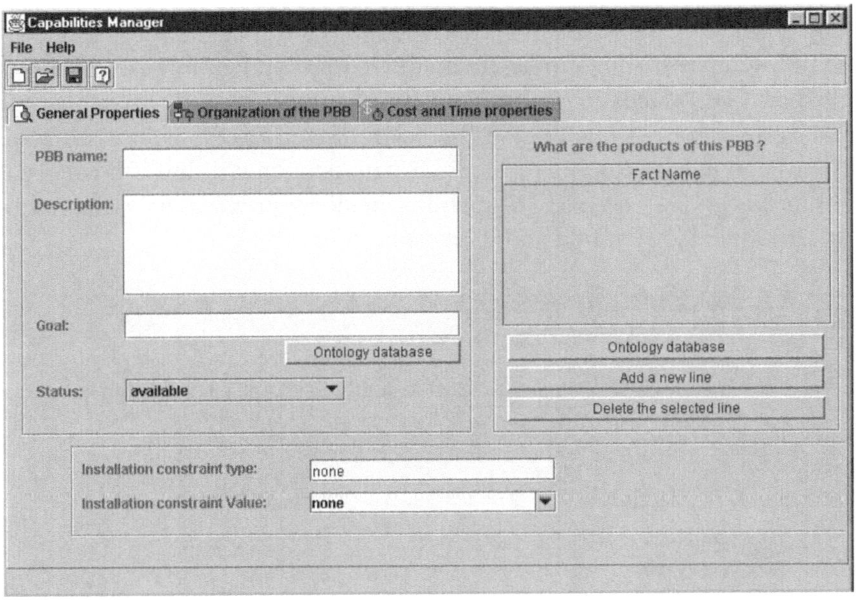

Fig. 7.3. General PBB properties

The PBB is typically triggered by the GCEN control procedures in situations where it seems to be a suitable choice. However, it is still up to the PBB to evaluate the situation and decide what and how to proceed. Inspired by the Belief-Desire-Intention (BDI) Theory (RAO and GEORGEFF, 1995 [80]) the PBB first analyzes the incoming SR, classifying the given situation into a number of belief states – the "Beliefs". Knowing the beliefs, the PBB then identifies what needs to be achieved in this situation – the "Desire". Finally it adopts a specific plan to achieve the desire – the "Intention". The BDI paradigm has proven successful in many cases in which expert approaches are to be captured and implemented in IT solutions. It is commonly used as the underpinning reasoning-model in many state-of-the-art agent systems and is supported by the Jack™ Agent System[46] used to implement GCEN.

In order to simplify the reasoning model conceptually and cater for the specific needs in GCEN the typical event driven meta-level reasoning used to decide upon belief states has been replaced by a simple decision tree model.

The tree is composed out of reasoning branches, beginning with a root branch "0". The user will assign a specific belief state to each branch and has to specify a set of logical tests which are carried out on the given service request. If the set of tests succeeded the PBB assumes to be in the specified belief state and adopts desires (goals) and intentions (plans) related to it. Having enacted the involved plans and achieved the goals the PBB proceeds to the follow-up reasoning branch to continue reasoning or invokes a follow-up PBB and terminates its invocation. In case the initial tests failed or the desires could not be achieved, the PBB will proceed to the next alternative branch to continue reasoning or the PBB's invocation will fail if no alternative branch is given. The GCEN capability manager provides a tool to graphically build the decision tree and its reasoning branches.

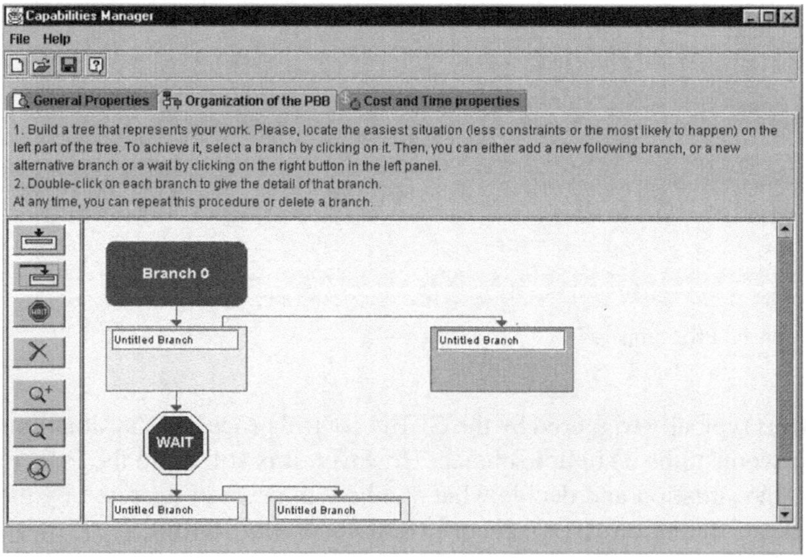

Fig. 7.4. The decision-tree-builder

46 JACK Intelligent Agents is a JAVA based agent system implementing intelligent BDI agents. It is provided by Agent Oriented Software Pty Ltd., 221 Bouverie Street, Carlton, Victoria, 3053, Australia. http://www.agent-software.com; info@agent-software.com.

The reasoning branch is the central element of the PBB's reasoning model. It holds the set of logical tests that identify the belief state ('Beliefs'). It defines what is to be achieved or which results are to be communicated to follow-up PBBs ('Desires'). Last but not least it holds the computational instructions ('Intentions') that have to be carried out successfully to achieve the desires.

For the definition of logical tests in the reasoning branches the user drags and drops operands from the ontology into the test section of the reasoning branch, building simple test language expressions. The system supports three different test types to identify the belief states:

a) **Necessity Test (n)**: a logical test that verifies an absolutely necessary condition. (Example: If the PBB has to arrange a delivery, the customer name has to be known ⇒ "n Belief('customer name') exists".) If this test evaluates to 'false' or one of its operands does not exist the reasoning branch cannot be considered.
b) **Disqualifier Test (d)**: a logical test that falsifies a condition that absolutely must not be 'true'. (Example: If the PBB has to arrange a delivery, the products may require not to be declared 'explosives' ⇒ "d Belief('additional product information') = 'explosives'".) If the operands exist and the test evaluates 'true' the reasoning branch cannot be considered.
c) **Precondition Test (p)**: a logical test that verifies a necessary condition for the enactment, where the assumption is made that the company is able to resolve this condition by invoking other capabilities. (Example: If the PBB has to arrange a delivery of products it needs to have all shipment papers prepared. If they are not prepared yet, the shipping department will have to prepare them first, which requires an additional SR to be posted. ⇒ "p Belief('shipment papers') = 'prepared'".) If the operands do not exist or the test fails, the PBB will post a SR, asking for the fulfillment of this precondition.

Next to the belief tests the user has to specify the result of the branch invocation in case it is enacted successfully. For this purpose the reasoning branches support a simple operand language to build up suitable status reports. Using the example above this might be as easy as " 'status of delivery' 'finished'".

The intention section of every reasoning branch allows the implementation of activities to be carried out. To achieve this, additional tests may be executed that do not interfere with the PBBs decision tree. Based on the results of any of the tests carried out 'fact name'-'fact value' pairs may be communicated to follow-up PBBs involving any desired operands. In addition an arbitrary number of operands can be handed over to so-called 'Reasoning Objects' that carry out numeric operations, modify or read databases, send e-mails or carry out any other programmable activities. (E.g. calculating and communicating the sum of operands: "Object('sum', Belief('warehousing costs'), Belief('finance costs'))".)

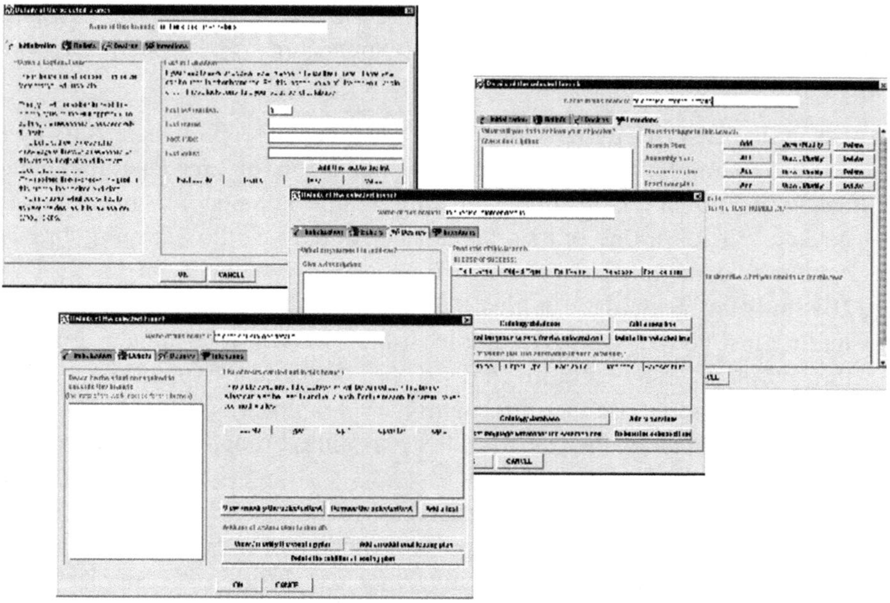

Fig. 7.5. Defining the reasoning branches

One of the most important activities triggered in a reasoning branch is the posting of "To Do's". In the above example 'arrange delivery' the employee might want to post "load truck" to his "To Do"-list. At this moment a number of additional facts are required and have to be included in the task description of the "To Do". Among them e.g. the order number, shipment paper number, customer name and address, list of products to be packed, the loading bay or the name of the forwarding agent and the like. Technically this is realized by posting a SR with the objective "post 'To Do'" to the GCEN system. The enactment of the "To Do" is regarded as the fulfillment of a precondition, which allows the employee to specify any information she or he wants to have handed over to her- or himself in the intention part of the branch that posts the "To Do".

The employee will finish to describe his capability by defining its cost and time properties. To estimate costs each PBB uses a simple activity based costing model (ABC-Model), in which the employee can define a number of fixed costs as well as a number of variable costs with their related cost rates and cost drivers (as arbitrary operand language expressions like: cost driver = "Belief('quantity')"). Most important for the management of resources and task scheduling is the availability of enactment time estimations. For this purpose every PBB implements analogous to the cost model also an activity based time model. The real enactment times and costs are later captured during process enactment. This allows a further calibration of both models.

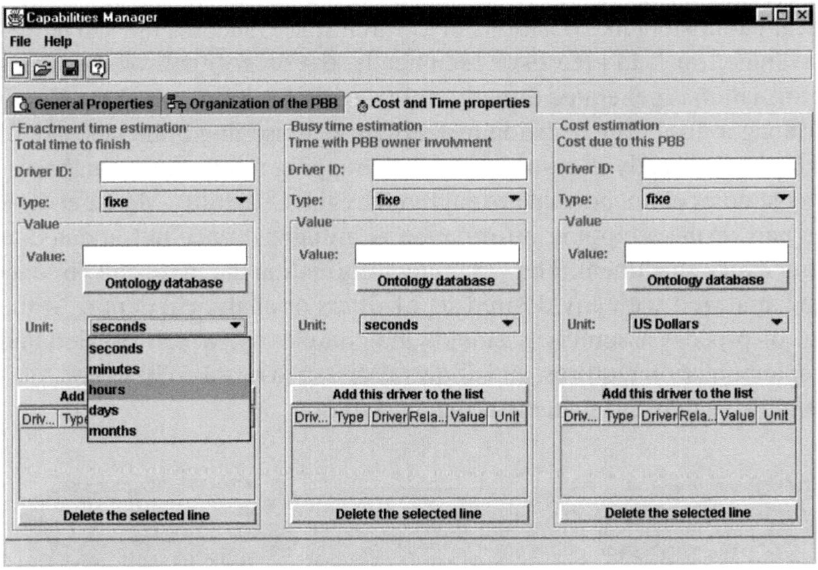

Fig. 7.6. Defining the PBBs cost and time properties

The decision tree together with its reasoning branches and all cost and time parameters represent the genetic set of a PBB. All test language expressions as well as all operand language expressions and static parameters are stored in the GCEN database. Once the PBB is invoked during runtime, these data sets are retrieved and interpreted. For this purpose, every PBB has a dedicated reasoning engine at its disposal. The reasoning engine will interpret the PBB's genetic set and with it will reason on the SR received. It will decide upon the PBB's applicability, will prepare the SR that the PBB communicates to the follow-up agent and will invoke any computational steps indicated.

7.2.3 The Wrapping Information

The conceptual approach to declarative process assembly, as introduced in chapter 7.3, adopts the use of so-called "Wrapping Information" from Landauer and Bellman's wrapping approach to declarative programming (s. Chap 7.3) (LANDAUER and BELLMAN, 1999 [48]). Wrapping information is machine interpretable meta information used to carry out the selections along the assembly process. It describes the PBB, its goals, where, whether, when and how to use it. Each GCEN PBB carries a number of data sets, which are stored in the wrapping information database. Among them are author, date and time of implementation, name and description, objective, expected output or results, context under which an application is appropriate, necessities, disqualifiers and preconditions but also empirical or numerical parameters like positions in a search space, success rates, average costs, average times and the like. Technically, the biggest part of the wrapping information is generated directly from the PBB's definition in the Capability Manager Tool. The decision tree structure, reasoning branches and all related tests are directly documented in the wrapping information database – so are most other elements that are captured by the Capability Manager. The empiric part of the wrapping information is initialized once but updated at least after every enactment. The wrapping information database can be searched and analyzed with any desired set of filters or analytical depth. At the moment of process assembly it shields the complexity, which the detailed PBB implementation requires, and allows decision to be taken based on relatively simple search and matchmaking algorithms.

Fig. 7.7. The wrapping information

The wrapping information guides the process assembly and with it, it defines the interplay of the available process building blocks. Thus the wrapping information represents an important part of the system's organization.

7.2.4 The Implementation of PBBs

Guided by the Capability Manager, the employee communicates her/his capability and defines its specific properties. After that the GCEN system has to build the agent that represents this PBB, generate its code, compile it and make it available in the system. The wrapping information has to be generated and stored into the wrapping information database.

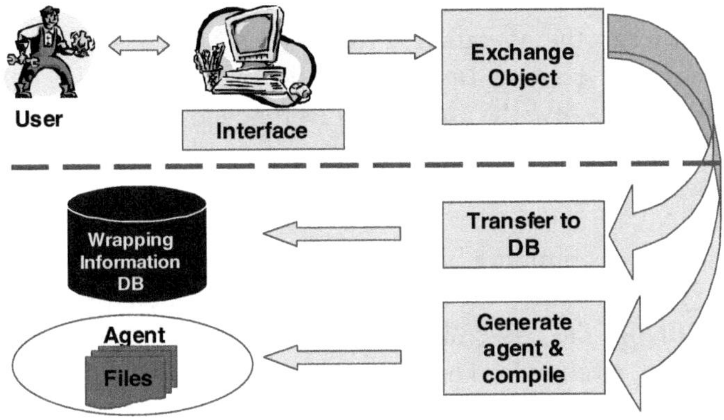

Fig. 7.8. The implementation of PBBs

Every time an existing capability should be modified again a new PBB agent is built carrying a different version number to facilitate a version control and ensure processes already assembled with the older versions still run to completion using the older versions during enactment. Again this requires the generation of code and wrapping information and again it is the system that finally builds and integrates this new PBB as a structural element.

The GCEN system thus constantly extends its own code libraries at runtime. New computational elements – new class files and object types – are generated while the system operates. The application itself, its physical software code, constantly adapts to the changing environment.

7.3 The Declarative Processing Environment

The 'Declarative Processing Environment' (DPE) represents the functional core of the GCEN system that realizes the declarative assembly of PBBs to a suitable response process. It is realized by a set of dedicated PBBs that mediate between the invocations of all other 'capability' PBBs. The four most important PBB types that form the DPE are:

a) The Carrier Agent (CA)

The carrier agent is the basic glue that binds any invocation of any other PBB to the process. Once about to terminate their invocation, every other PBB searches for the most up-to-date and valid carrier agent and invokes it by handing over a Service Resquest (SR). This SR contains on one side the documentation of what has been done and on the other side includes possibly several requests for preconditions to be fulfilled. The CA documents the invocation results of the previous PBB as well as all requests for further preconditions to be satisfied to the so-called 'Project Fact Sheet' (PFS) that accompanies every project. Having finished the documentation, the CA invokes follow-up PBBs to resolve the preconditions for each of which a new subproject is opened. Thus the CA carries the project from stage to stage.

b) The Assembler

The 'Assembler' is a process building block that is exclusively invoked by the CA. At any given point, at which the CA does not know which follow-up PBB to invoke, it will invoke the 'Assembler' and handover a SR that describes the given problem and its context. The 'Assembler' will analyze the situation and decide upon a possible PBB based on the existing sets of wrapping information. Thereafter it will invoke the CA again and propose its selection. To carry out the matchmaking the 'Assembler' was programmed to address a number of pluggable search and analysis tools, all of which return a weighted result table. The 'Assembler' merges the tables and carries out a random but weighted draw – it thus does not follow a deterministic but rather a probabilistic algorithm to carry out its selections.

c) The SUII Agents (Standard User Input Interface Agents)

Unlike all other GCEN PBBs, the SUII agents do not terminate their invocation right after posting a service request to the follow-up CA. The SUII agent remains up and running as long as the user to whom it belongs is logged on. Each employee using the system has her/his own SUII agent, which provides an access point to the GCEN system: a generic interface to build formal SRs and post them to the system. The generic interface offers drag-and-drop support to generate service requests quickly and easily out of ontological expressions. It is a "standard" input interface as it is not designed to cater for a particular type of SRs – customized interfaces supporting a highly efficient data input for dedicated SR types still have to be designed. The SUII agents are regarded as a part of the infrastructure that implements the DPE as they represent the input points that capture any standard stimuli to the system upon which a process will be assembled.

d) The Scheduler Agents

Scheduler agents manage resource constraints. For each constrained resource, which might be used by the GCEN system, there exists one dedicated 'Scheduler' agent. The most important constrained resources are the employees themselves: they have a limited capacity to enact "To Do's", as well as they have Saturdays, Sundays, public holidays, annual leave or simply meetings that may occupy their time. Each employee thus is represented by its own 'Scheduler' agent. It handles SRs with the objective "post 'To Do'". The agent analyzes time constraints of both employee and task, makes reservations, handles the SR or rejects it if the task could not be scheduled. During the process enactment the 'Scheduler' agent presents the "To Do's" to the employees and guides them through the fulfillment of their tasks. This way enactment results are captured, documented and considered throughout the process. The Scheduler agents, hence, are very special PBBs that do not represent an employee's capability as such – they rather belong to the infrastructure that builds up the DPE.

The process assembly is organized in recursive assembly cycles that subsequently select suitable PBBs. At the beginning of the assembly cycle there is a well-conditioned **service request** (SR) – the initial stimulus to trigger the cycle. The second step in the assembly cycle is the **global selection** proposing a PBB that is most likely to be able to fulfill the SR. The selection is initiated by the CA and carried out by the 'Assembler'. The third step of the assembly cycle involves the invocation of the proposed PBB and its **local verification**. The PBB will reason upon the given situation and evaluate its applicability (-> decision tree, reasoning branches and reasoning engine; s. Chap 7.2.2). The evaluation will lead to one of the following results: rejection, conditional acceptance or unconditional/total acceptance.

In the case that the process building block rejects the service request a new process building block has to be found by the selection mechanism. In the case of a total acceptance the process building block assumes that it will fully satisfy the service requests once it is enacted. If the PBB found itself only conditionally applicable it has to have one or more preconditions to be fulfilled first, in order to be enacted successfully. Thus the PBB will post new SRs for each precondition[47]. In this case a local goal decomposition takes place leading to new assembly cycles.

---

[47] A PBB that requires several preconditions to be fulfilled prior to its successful enactment will more precisely post only one formal SR. This SR includes or communicates several objectives that need to be satisfied.

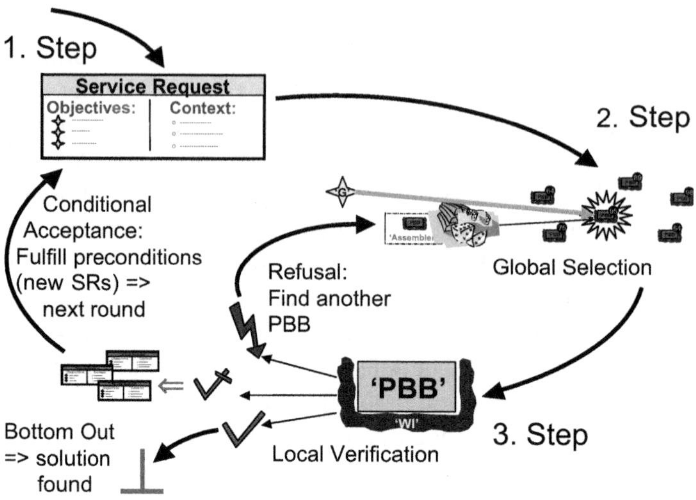

Fig. 7.9. The assembly cycle

The process assembly will deliver a process in the form of a hierarchical tree of PBBs. These tree sets in a sense encode or represent a particular solution. They are used to guide the algorithms to evaluate and enact processes as well as they are used for representation and documentation purposes.

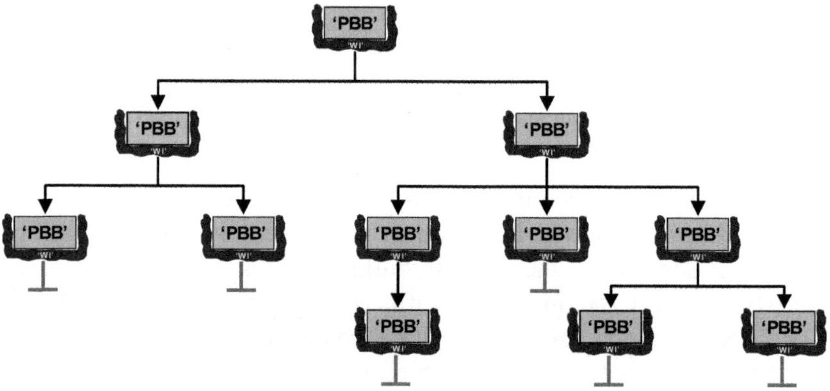

Fig. 7.10. Tree sets – the macroscopic process representation

Any assembly is constrained to a maximum assembly time in order to ensure the system's responsiveness. Within this time, zero, one or several solutions may be found. If the system was unable to find a solution, the SR was outside the system's cognitive domain. In case several solutions have been worked out they can be evaluated according to given profit, cost and time considerations. Finally the DPE will propose the most suitable solution for further enactment.

When a decision has been made to enact a particular solution the GCEN system will continue with the so-called reservation run. Once all resources have been successfully reserved, the enactment will be triggered.

Phase I:	Phase II:	Phase III:
Assembly	Reservation Run	Enactment
Service Requests are analyzed and appropriate PBBs invoked. Local verification is carried out and further SR are posted as needed.	A assembled solution has been selected. Every PBB is invoked again to allow it to make any necessary reservations for resources it requires.	The assembled process is enacted. Every PBB is invoked and any necessary enactment actions are executed.
First sets of facts are generated and recorded in the PFS based on results of logical tests and estimations	A second time sets of facts are generated and recorded based on results of logical tests and estimations	During Enactment the real facts are generated and recorded to the PFS

Fig. 7.11. Assembly, reservation run and enactment

7.4 The Autopoietic Framework

The Autopoietic Framework (AF) is the second core element of the GCEN system. Again, it is implemented by a set of dedicated PBBs. But in contrast to the DPE PBBs, the AF PBBs interact with the DPE as normal building blocks that allow the instantiation of processes to evaluate previously enacted company processes, refine the system's organization and build new PBBs (structure).

7.4.1 The Evaluation Mechanisms

a) Readily Available Parameters

The cost and time model as well as the observation of beginning and completion times of tasks provide a number of stable estimations, among them:

- enactment time per PBB: $t_{en}^{(i)}$
- enactment costs per PBB, including costs for resources: $C_{en}^{(i)}$
- total enactment time per process: $t_{en}^{(tot)}$
- total enactment costs per process: $C_{en}^{(tot)}$

Once the services have been sold and correctly billed externally or internally the following parameters can be estimated:

- total revenue per process: $R^{(tot)}$
- profit per process[48]: $P^{(tot)}$

This is the set of basic parameters available in the GCEN system. Any further evaluations are based on these six parameters.

Without doubt, there are a number of other points that should be considered in comprehensive evaluation paradigms. The most important among them are: customer satisfaction, employee motivation and quality of services or products. Up to now no approach has been implemented to consider these – conceptually more difficult and partly intangible – parameters.

b) Learning Curve Analysis

Regarding the enactment costs per PBB, a continuous decay over time can be expected, as the employee enacting the task will benefit from learning effects. The cost development over time should follow the so-called learning curve, which was introduced by Wright as follows (WRIGHT, 1936 [122]):

$$C_{en,estim}^{(i)}(x) = a^{(i)} \cdot x^{b^{(i)}} \qquad [7.1]$$

Where x is the number of enactments, c the expected costs at the xth enactment, $a^{(i)}$ is the cost of the first enactment and $b^{(i)}$ the rate of improvement ($b^{(i)} < 0$ for typical learning curves).

Competitive advantages can only be gained if the enterprise learns faster than its competition in terms of time units. Thus for strategic reasons it is not the number of enactments which is of interest but rather the time within which learning effects can be achieved. Furthermore, we assume that the costs may contain fixed elements that show no learning effect. Thus we propose:

$$C_{en,estim}^{(i)}(t) = a^{(i)} \cdot t^{b^{(i)}} + c^{(i)} \qquad [7.2]$$

The parameters $a^{(i)}$, $b^{(i)}$ and $c^{(i)}$ can be obtained easily from statistical data by simple regression analysis. They characterize the task encapsulated in the PBB in terms of their 'learn-ability' that was achieved by the employee.

Assuming the validity of Frederick W. Taylor's theoretical approach (TAYLOR, 1911 [106]), on the one hand every arbitrary complex task can be broken down into a series of simple activities. Simplicity on the other hand implies 'characterize-ability' and comparability in that similar activities, to

48 The estimation for the achieved profit per enacted process is given by: total revenue per process minus total enactment costs per process. Note that the profit estimation is just as good as the cost estimation.

a certain extent, show comparable learning curve characteristics. Also if in a diluted form, the same should still be valid on PBB level: PBB clusters with similar characteristics should show similar learning curve parameters. With the help of pattern recognition tools, like neuronal networks, the wrapping information of PBBs and their learning curve parameters can be analyzed and trained. This allows later to retrieve internal reference values for the enactment costs to any given set of wrapping information.

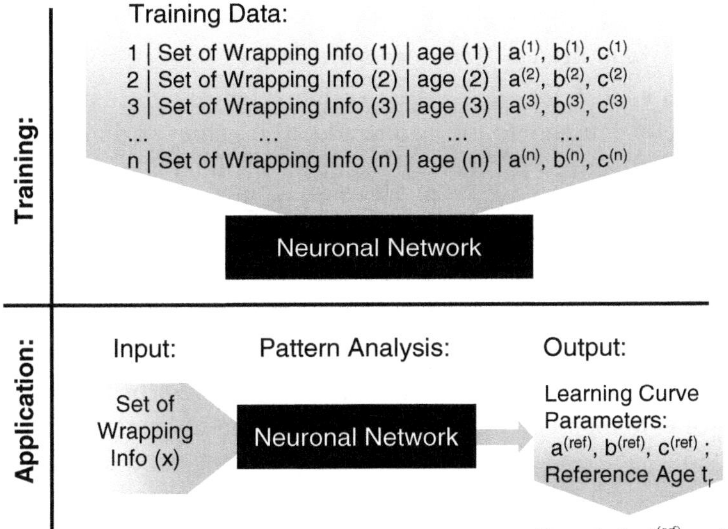

$$C_{ref}^{(i)} = a^{(ref)} \cdot t_r^{b^{(ref)}} + c^{(ref)}$$

Fig. 7.12. Learning curve analysis and pattern recognition

At any given process enactment, the contribution a single PBB – or more precisely the employee providing the PBB – made can be estimated by comparing the real enactment costs to the reference values obtained from the learning curve analysis. As the company has no access to any other capabilities/PBBs than those brought in by its employees this estimation may be regarded fair even so it is not taking industry standards into account (self-reference).

It is worth mentioning some of the conceptual shortcomings or critical points of this approach:

– Good results can be expected for low-level capabilities whereas for high-level capabilities inaccurate reference values may be obtained as a real comparability may not be given.

- The approach requires a sound statistical base without which the accuracy of obtained values may suffer.
- The large amount of wrapping information available per PBB may cause significant noise slowing down or hindering effective learning of the pattern recognition tools.
- Tool and learning strategies used may have a significant influence on the predicted results.
- The reference values are obtained fully self-referenced. This may occasionally require significant corrections by the management.

c) Estimating the 'Value Added' Achieved by a PBB

As described in chapter 6.5.1 there is a strong relation between industry standard costs an activity causes and the 'value added' it achieves. In order to obtain reasonable estimations for the 'value added' the internal cost values thus need to be calibrated to the outside world. From a microeconomic perspective, the profits (or losses) achieved by the enactment of a task is exactly the difference between the 'value added' it achieves and the internal costs it generated. The sales department obviously establishes or even enforces a calibration by adjusting its sales price to the market requirements.

Allocating the profit (or vice versa losses) a process achieved (or caused) to the PBBs according to their contribution (s. discussion in b)) and deducting further the materials and goods employed would provide a valid estimation to the 'value added' each of them achieved. Taken as an isolated value any estimation is vague and inaccurate, applying however the same tools for 'value added' as previously for cost estimations (regression analysis and pattern recognition) does again lead to reference values for 'value added' to any given set of wrapping information.

Note:
- The described approach builds on the effect that a given PBB contributes under numerous different circumstances and thus contextual effects (e.g. from dominant neighbor PBBs) are eliminated. Thus again the estimation is more stable for low-level PBBs that are used in many different processes, whereas high-level PBBs may lead to problematic results.
- Not furthermore, that profit alone does not always allow a stable calibration. Other elements, such as market segmentation, market structure and market share might have to be considered, too.

d) Profit Achieved per PBB

Given the reference value for the 'value added' a PBB with given wrapping information typically achieves, an estimation for the profit it was able to obtain at a given enactment is easily calculated: the profit equals the 'value added' (reference value) minus all enactment costs (excluding goods and materials employed) that resulted at the given enactment.

Note that this profit estimation is already cleared from profits or losses caused by price fluctuations in the goods and materials employed.

7.4.2 Refinement of the System's Organization

As introduced in chapter 4.1.4 the system's organization defines the functional interplay of its structural components. In other words, it defines the processes that are enacted at a given situation. Transferred to the GCEN system the following three elements form its organization:

- the wrapping information
- the Carrier Agent (CA) and
- the Assembler.

Autopoietic processes to refine the system's organization adapt and fine-tune the sets of wrapping information as well as the behavior of CA and Assembler with the help of the existing evaluation results. So is, for example, the success of proposals made by the Assembler evaluated and the way it merges the result tables from different search engines adjusted to optimize its decisions. Wrapping information like average times, costs, success rates, positions in search spaces and the like are updated with the values retrieved from the latest enactments.

Example: The Brain Model
The GCEN system implements a simple search space that allows to map historic data about the success of selections under different circumstances. For this purpose every PBB and all of its preconditions (or possible SR types) it will post are captured as mass points in a virtual three-dimensional search space. Their movements follow a simple kinetic model: After the enactment of a process the success of all selections that chained the PBBs will be evaluated in terms of the profit a PBB-pair achieved. Each pair of PBB ending (SR of the first PBB) and PBB core (second PBB) will receive a virtual impulse moving them closer or separating them according to the profit or losses this link caused. The model builds on a well-balanced adjustment of dissipative friction, conservative potentials and induced impulses, which keeps its density stable and prevents PBBs from clocking together. The PBBs weights (as well as the weights of it's endings) and thus its tendency to move depends on the frequency of its usage – heavily used PBBs aggregate virtual mass and thus occupy stable strategic positions.

The model realizes memorization in terms of the positions cores and endings occupy. After several runs, a PBB ending (representing a particular type of SR) will move into an area where it is surrounded by PBB cores that tend to handle its SRs successfully, whereas the PBB cores will move into areas in which SRs exists that they can handle. During process assembly, the Assembler uses a tool, the 'Brain Model Search Engine', to find PBBs close to a given service request. This tool builds up and returns a result table based on the distances between SR (PBB ending) and the cores of PBB candidates. Distances thus translate into probabilities for selection.

All the positions of core and endings are part of a PBB's wrapping information, which are constantly updated by the Autopoietic Framework (AF).

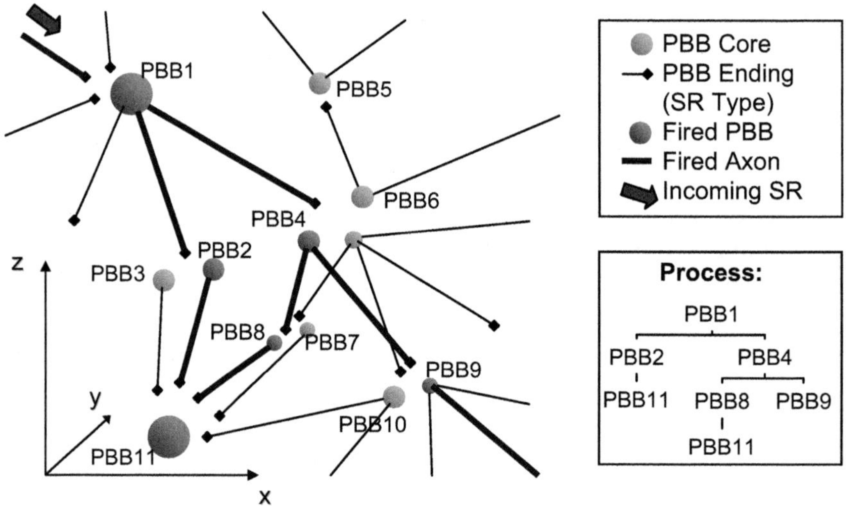

Fig. 7.13. Stimuli propagation in the brain model

7.4.3 Refinement of the System's Structure

The PBBs are the system's structural elements, thus it is up to the Autopoietic Framework to constantly introduce new PBBs and replace old ones. For this purpose the GCEN prototype was prepared for three dedicated autopoietic processes:

a) Creation

The creation processes have been discussed and described in chapter 7.2. They enable the system to capture new atomic PBBs that an employee is willing to offer to the system. The creation processes will guide the employee and help him to describe his capability. Together with the employee a set of wrapping information will be generated. Thereafter, the PBB's code is generated, compiled and registered at the system.

b) Internalization

Internalization processes analyze successful enactments and identify highly successful tree-sets of PBBs. A new compound PBB, representing such a tree-set, is then generated which will be available for future assemblies. This way the system internalizes best practice. Internalized PBBs significantly reduce assembly time and increase the probability of successful solutions.

c) Rethinking

The rethinking processes analyze those PBBs that show a high utilization and try to rebuild them using other, probably newer and more effective PBBs. Rethinking processes ensure that new and effective PBBs find access to and improve known best practices. They also replace important but inefficient PBBs – that might not have been depreciated by reinforcement learning due to their high utilization – by simply populating the PBB pool with better solutions.

The automated generation of new PBBs in the GCEN system is realized through the conceptual similarity of macroscopic process representation (of enterprise processes) and the representation of microscopic reasoning in the PBBs (the PBBs genetic sets): both are tree-sets. Given the decision trees and reasoning branches of several PBBs and the way these PBBs are supposed to be chained (macroscopic process) a new decision tree including all the original branches can easily be generated forming the genetic set of a new and executable PBB that integrates several capabilities.

7.5 System Architecture

As with every IT application, the system's architecture has to support a long list of requirements and thus it is crucial for any successful implementation. Some of the most important considerations that finally lead to the chosen architecture of the GCEN system are:

- The system has to reach all involved employees right at their desktops.
- The system has to support the implementation of PBBs with autonomous reasoning abilities.
- The code base has to be open during runtime to allow continuous integration of new structural elements (program code).

The selected architecture builds on Jack™, a state-of-the-art intelligent agent system provided by Agent Oriented Software Pty Ltd.. It supports a full network distribution using so-called 'Portals' that run on an arbitrary number of host machines each of which holds a TCP/IP connection to the rest of the system. The core facility, which allows the demand driven assembly of processes, is implemented as a dynamic agent community living in the system's portals. It is supported by the **GCEN file server** – providing the code base –, the **GCEN database server** – holding several databases including the wrapping database – and the **GCEN PFS portal** – on which all project and process documentations, the PFSs, are kept.

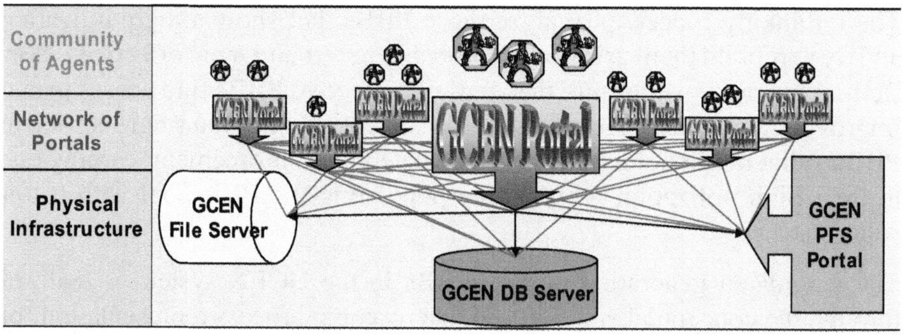

Fig. 7.14. The GCEN architecture – base layer

On top of this layer, all the described GCEN entities are implemented:

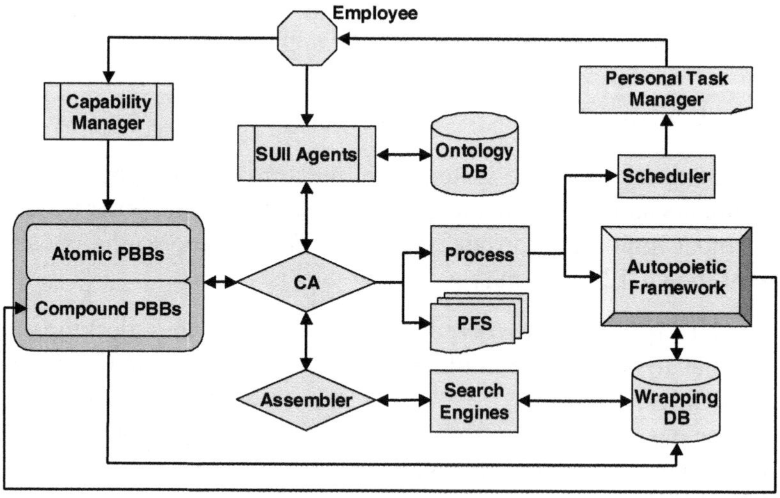

Fig. 7.15. The GCEN architecture – application layer

7.6 Characteristics of the Running System

Triggered by a stimulus, the SR, the GCEN system tries to assemble a suitable response process. It checks the availability of resources and reserves them as needed. Right away it schedules "To Do's" for all involved employees who can then instantly start to enact their tasks. The employees are relieved of any process-related coordination efforts and can concentrate on the fulfillment of tasks.

Processes are generated and tailored to a specific situation, not to establish the enterprise's standard approaches. The traditionally most feared 'Exception' has become the 'Rule'! If an assembled process cannot be executed as proposed the system does conceptually exactly what it does in any other situation: it assembles a new solution, removes previous "To Do's" and resource reservations and places new ones. This is a computational effort but does not bind the company's workforce.

The assembly and proposal as well as enactment of a response process to a given stimulus is a cognitive act. This may not seem obvious in the first place but should get plausible if the consequences and applications are considered that it might have.

Cognition typically expresses itself in the ability to perceive and interpret stimuli, especially if these do not have a standardized and well-conditioned format. Given a complex stimuli pattern like the sentence "Please deliver 100 packs of UHT milk to Superstore Ltd. on 01.08.2005" it would be perceived and interpreted by any involved employee in that "deliver" indicates what, "Superstore Ltd." indicates to whom and "01.08.2005" indicates up to when it needs to be done – this is the interpretation her/his cognitive ability has sorted out as meaningful. Given the same sentence to GCEN, first a translation into a service request (SR) would be necessary. This could be easily realized by algorithms that utilize the system's ontology, but would in general not lead to a definite translation but rather to a number of possible interpretations, such as: (s. Fig. 7.16)

Out of these SRs however the system will only be able to successfully assemble a response process for SR version 2. Only it represents a meaningful interpretation of the given challenge and would be adopted to guide any enactment. The described perception and interpretation of course may lead to irritations or cause unexpected system behavior – but this is what may identically happen even if humans perceive and interpret. The GCEN system thus realizes process-based, intelligent perception as discussed in chapter 4.6.1.

"Please deliver 100 packs of UHT milk to Superstore Ltd. on 01.08.2005"

SR version 1	
objective	deliver
product	packs UHT
quantity	100
customer	milk to Superstore
due date	01.08.2005

SR version 2	
objective	deliver
product	UHT milk
quantity	100 packs
customer	Superstore Ltd.
due date	01.08.2005

SR version 3	
objective	deliver
product	UHT milk
quantity	01.08.2005
customer	Superstore Ltd.
due date	100

SR version 4	
objective	deliver
product	Please
quantity	100 packs
customer	Superstore Ltd.
due date	01.08.2005

Fig. 7.16. Possible results of automated translations

The PBBs are the system's structural elements. They interact with each other and with the system's environment (the employees), propagating stimuli and realizing processes. The community of PBB agents sets up the space in which the system exists and can be perturbed. Yet they are merely communications. Communications which have been articulated by the employees and manifested in the system. They are not physical parts, hardware components or any other aggregation of physical matter. The virtual structural elements are given only as software entities. In a sense, thus, the PBBs may loosely be interpreted as a conceptualization of Luhmann's 'communications' – the constituting structural elements of social systems. A conceptualization that clearly defines these elements, their interactions and the space in which they exist.

Chapter 8 Conclusion

A Brief Summary of Our Findings

In today's competitive business environment every enterprise is torn by

- the need to be extremely precise and live with smallest margins,
- the need to handle complexity in products and information but also in the steps that have to be taken as well as the decisions that need to be made and
- the need to be dynamic facing a quickly changing environment and diminishing market life-cycle times.

These three mutually opposing forces are neither phenomenologically new nor unknown to the world of business and engineering management – what is new and extremely challenging is the rapid development in which these forces gain strength.

This requires the enterprises to interact and behave, using an advanced conceptual paradigm: **Intelligence**. The enterprise has to develop cognitive abilities and build up a framework that allows it to generate highly effective, intelligent behavior: **Enterprise Knowledge**. Being distinct from their human counterparts, knowledge and intelligence on enterprise level require more than organizational learning, data warehouses and knowledge bases or traditional Knowledge Management. They need far-reaching physical and organizational changes that prepare the company's organization and structure to support the development of knowledge and intelligence. Processes have to be understood as one-time instances and the practice of dynamic, declarative processing has to replace today's rigid process landscapes. The employee has to be integrated as a real knowledge worker, a functionally autonomous entity providing her/his own abilities to solve problems and seek for solutions. Last but not least the enterprise's IT infrastructure needs to be understood and used as the physical framework that sets up and realizes its processes and drives its behavior. It has to form the enterprise's central nervous system, which enables its responsiveness. Future IT systems will replace traditional ERP approaches and provide solutions far beyond their scope. Finally it is the IT systems that have to realize declarative processing, assembling tailor-made processes right to the requirements perceived in a given situation. Handling and managing the network of communications moves to the center of interest, as it will define the enterprise's cognitive domain, its responsiveness, core competences and ability to generate 'value added'.

Along the theoretical framework built up, an IT prototype, called GCEN, has been implemented to visualize its practical application and consequences. The GCEN system illustrates how the declarative assembly of processes out of generic 'process building blocks' (PBBs) might be technically realized. Furthermore, it provides an environment to experiment with and build examples for real PBBs, each capturing and representing an employee's capability. With the system being 'open' at runtime, continuously changing and adapting its underlying code libraries – rebuilding its structure and organization – it establishes the foundation for intelligent learning and puts a prominent paradigm right into practice: Autopoiesis. Last but not least, GCEN realizes and guides meaningful response behavior and content-based perception, demonstrating cognitive phenomena on enterprise level.

Like any prototypical implementation, the GCEN system does have its limitations. For the application in the 'Real World', the system's use, the posting of Service Requests and the formulation of capabilities (generating PBBs) have to be easy enough to be handled by the employees themselves. Its response times (times to finish a process proposal) have to be short, data security and consistency (in case of network or system failures) have to be high and the system has to run stably over long periods of time. All these are points that have not been emphasized at the implementation of the prototype.

This work's main contributions are the new understanding of the enterprise as a natural system as well as the theoretical framework that attempts to explain the emergence of enterprise intelligence and enterprise knowledge. The enterprise's network of processes moves to the center of interest. Unique, however, is that processes in intelligent enterprises are regarded as one-time instances, assembled on-the-fly and tailored to a specific situation. The key to knowledge and intelligence thus is the self-generation and self-organization of building blocks to a process, enabling intelligent behavior (Autopoiesis, Declarative Processing). Both, the developed theoretical framework and the implemented software prototype GCEN, have inspired ongoing research, applying them to several specific Industrial Engineering problems (CHEN, ZHANG, GUPTER and TSENG, (2001) [17]; TSENG, CHEN, GUPTER and WANG (2002) [111]).

Yet, there are uncountable questions still waiting to be answered. Questions like "Is there a finite set of PBBs (capabilities) that would allow an enterprise to enact the full variety of processes it needs?", "How high or low level are capabilities to be best expressed in PBBs?", "What are the best approaches to rebuilding structural components?", "How are irritations perceived and how do these perceptions influence structure and organization?" or "How will managers and employees handle their new roles in a constantly changing process landscape?". These and many other questions will have to be answered in the years to come.

List of Figures and Tables

Bibliography

[1] Allport, F.H.: *Social psychology.* Houghton Mifflin, Boston (1922).

[2] APA (American Psychological Association): *Intelligence: Known and Unknowns.* Science Directorate, Washington (1995).

[3] Aristotle: *The History of Animals.* Book VIII, Corpus Aristotelicum (350B.C). [Translation e.g. at: http://classics.mit.edu/Aristotle/history_anim.8.viii.html (7/15/01) by D'Arcy Wentworth Thompson.]

[4] Ashby, W. R.: *Design for a Brain: The Origin of Adaptive Behavior.* Chapman and Hall, 2nd ed., London (1960).

[5] Banzhaf, W. et al. : *Genetic Programming.* Morgan Kaufmann Publishers, San Franzisco (1998).

[6] Barnett, M. R.: *Fundamental Particles And Interactions.* http://www.cpepweb.org (7/27/01) Contemporary Physics Education Project and Particle Data Group of Lawrence Berkeley National Laboratory, Berkeley (2000).

[7] Baumoel, U.: *Die (R-)Evolution des Informationsmanagement.* Gabler, Wiesbaden (1998).

[8] Bertalanffy, von, L.: *General System Theory.* George Braziller, New York (1968).

[9] Bertels, T.: *Organizational Epistemology by Georg von Krogh and Johan Roos – Reviewed.* International Journal of Organizational Analysis, Vol.5, No.2 (1997), pp. 205-207.

[10] Bontis, N. : *Managing Organizational Knowledge by Diagnosing Intellectual Capital: Framing and advancing the state of the field.* International Journal of Technology Management, 18, 5/6/7/8, 433-462.

[11] Bontis, N. et al.: *The Knowledge Toolbox: A review of the tools available to measure and manage intangible resources.* European Management Journal, 17, 4, (1999), pp 391-402.

[12] Bontis, N. : *There's a price in your head: Managing intellectual capital strategically.* Business Quarterly, (1996), pp 40-47.

[13] Botez, A.: *Michael Polanyi and Lucian Blaga as Philosophers of Knowledge.* http://www.bu.edu/wcp/Papers/Comp/CompBote.htm (12/16/01) Twentieth World Congress of Philosophy, Boston University, Boston (1998).

[14] Bullinger, H.-J., Woerner, K. and Prieto, J.: *Wissensmanagement heute: Daten, Fakten,Trends.* Fraunhofer (IAO) Institut für Arbeitswirtschaft und Organisation, Stuttgart (1997).

[15] Campbell, D. T.: *Evolutionary Epistemology*. In: Schilpp, P. A. (ed.): 'The philosophy of Karl R. Popper.', Open Court LaSalle, Illinois (1974), pp. 412-463.

[16] Chaisson, E. and McMillan, S.: *Astronomy Today*. Prentice-Hall, Third Edition, New Jersey (1998).

[17] Chen, Y.J., Zhang, M., Gupter, A., Tseng M. M.: *Autopoietic Approach for Production Management in Mass Customization*. 2001 Mass Customization and Personalization, Hong Kong (2001).

[18] Cornman, J. W. et. al: *Philosophical Problems and Arguments: An Introduction*. Hackett, Third Edition, Indianapolis (1987).

[19] Darwin, C. : *On the Origin of Species by Means of Natural Selection or the Preservation of Favoured Races in the Struggle for Life*. Murray, London (1859).

[20] Davenport, T. H. and Smith, D. E.: *Managing Knowledge in Professional Service Firms*. In: The Knowledge Management Yearbook 2000-2001, Butterworth Heinemann, Boston (2000), pp. 284-299.

[21] Deleuze, G.: *Differenz und Wiederholung*. Trans. by Vogl, J., Munich, 2nd edition (1997).

[22] Derrida, J.: *Positionen. Gespräche mit Henri Ronse, Julia Kristeva, Jean-Louis Houdebine, Guy Scarpetta*. Engelmann, P. (ed.), Böhlau, Graz/Wien (1986).

[23] Douglas, M.: *How institutions think*. Syracuse University Press, New York (1986).

[24] Drucker, P. : *Concept of the cooperation*. Transaction Publishers, New Brunswick (1993), pp 230-240.

[25] Drucker, P. : *Knowledge-Worker Productivity: The Biggest Challenge*. California Management Review, Vol. 41, No. 2, (1999) pp. 79-94.

[26] Drucker, P. : *Management: Tasks, Responsibilities, Practices*. Harper & Row, New York (1985).

[27] Drucker, P. : *The Post-Capitalist Society*. HarperBusiness, New York (1994).

[28] Eberl, P. : *Die Generierung des organisationalen Wissens als selbstreferentieller Prozess*. http://www.wiwiss.fu-berlin.de/w3/w3schrey/komwis/Beitraege/eberl.htm, Verband der Hochschullehrer für Betriebswirtschaft e.V., Kommission Wissenschaftstheorie (Internet 9/7/01).

[29] Foster, R.: *Innovation: The Attacker's Advantage*. Summit Books, New York (1986).

[30] Gabler: *Wirtschafts-Lexikon*. Vol. 8, 8th edition, Gabler Verlag, Wiesbaden (1995).

[31] Gladstone, B. and Kawalek, J. : *Problems on the Virtual Shop Floor: A Case Study*. In: The Knowledge Management Yearbook 2000-2001, Butterworth Heinemann, Boston (2000), pp. 242-250.

[32] Gordon, J.: *The Whole Enchilada: Intellectual Capital and You*. In: The Knowledge Management Yearbook 2000-2001, Butterworth Heinemann, Boston (2000), pp. 157-168.

[33] Greenwood, W. : *Harnessing Individual Brilliance for Team Creation – The Six C's of the Knowledge Supply Chain.* http://www.brint.com/papers/submit/greenwoo.htm, Online Collaboration Conference, Berlin (1998).

[34] Hansen, M. T., Nohria, N. and Tierney, T.: *What's Your Strategy for Managing Knowledge*, In: The Knowledge Management Yearbook 2000-2001, Butterworth Heinemann, Boston (2000), pp. 55-69.

[35] Hatvany, J. : *Intelligence and Cooperation in Heterarchic Manufacturing Systems.* Robotics and Computer Integrated Manufacturing, Vol. 2, No. 2 (1985), pp.101-104.

[36] Haubold, H. and Mathai, A. M.: *Structure of the Universe.* In Trigg, G. L.: Encyclopedia of Applied Physics, Vol. 23, WILEY-VCH Verlag, Weinheim (1998), pp. 47-81.

[37] Heidegger, M.: *Identität und Differenz.* Neske, Pfullingen (1957).

[38] Heisig, P.: *Knowledge Management and Kaizen.* In the proceedings of 2nd International EuroCINet Conference on Continous Improvement: from idea to reality, Enschede (1998).

[39] Jiao, J. and M.M. Tseng : *An Information Modeling Framework for Product Families to Support Mass Customization Production*, Annals of the CIRP , Vol 48/1 (1999), pp. 93-98.

[40] Johnson, A. R.: *Competitive Intelligence and Competitor Analysis as Knowledge Management Applications.* In: The Knowledge Management Yearbook 2000-2001, Butterworth Heinemann, Boston (2000), pp. 85-97.

[41] Kinni, T. B.: *With an Eye to the past: Transmitting the Corporate Memory*, In: The Knowledge Management Yearbook 2000-2001, Butterworth Heinemann, Boston (2000), pp. 208-213.

[42] Kowalski, R.: *Algorithm = logic + control.* Communications of the ACM (Association for Computing Machinery, New York), 22(7), (1979), pp. 424-436.

[43] Koza, J. R. : *Genetic Programming: On the Programming of Computers by Natural Selection.* MIT Press, Cambridge (1992).

[44] von Krogh, G. and Roos, J.: *Organizational Epistemology.* St. Martin's Press, New York (1995)

[45] Kuhn, H. and Waser, J. : *Selbstorganisation der Materie und Evolution früher Formen des Lebens.* In Biophysik, Springer Verlag, Berlin (1982), pp 859–907.

[46] Kunzmann, P., Burkard, F.-P. and Wiedmann, F.: *dtv-Atlas Philosophie.* dtv-Verlag, Munich (1991).

[47] Kutschera, F. von: *Grundfragen der Erkenntnistheorie.* De Gruyter, Berlin (1982).

[48] Landauer, C. and Bellman, K.: *Agent-based information infrastructure.* In Proceedings of Agent-Oriented Information Systems 1999 (AOIS-99).

[49] Levinthal, D. A. and March, J. G.: *The myopia of learning.* Strategic Management Journal, 14 (1991), pp. 95-112.

[50] Lorenz, K.: *Behind the Mirror.* Methuen, London (1977).

[51] Luhmann, N.: *Die Gesellschaft der Gesellschaft.* Suhrkamp, Frankfurt am Main (1997).

[52] Luhmann, N.: *Die Wissenschaft der Gesellschaft.* Suhrkamp, Frankfurt am Main (1992).

[53] Luhmann, N.: *Soziale Systeme. Grundriß einer allgemeinen Theorie.* Suhrkamp, Frankfurt am Main (1984).

[54] Luhmann, N.: *The Autopoiesis of Social Systems.* in Geyer, F., and van der Zouwen, J. (eds.): Sociocybernetic Paradoxes, SAGE Publications, London (1986).

[55] Malhotra, Y.: *From Information Management to Knowledge Management: Beyond the "Hi-Tech Hidebound" Systems.* In Srikantaiah, T. K., Koenig M. E.D.: Knowledge Management for the Information Professional, ASIS, Medford (1999), pp. 37-61.

[56] Mankiw, G. : *Principles of Economics.* The Dryden Press, Fort Worth (1998), p. 363-377.

[57] March, J. G.: *The Pursuit of Intelligence in Organizations.* In: Lant, T. and Shapira, Z.: Organizational Cognition, Computation and Interpretation, LEA Publishers, Mahwah (2001), pp. 61-72.

[58] Maturana, H. and Varela, F.: *Autopoiesis and Cognition: The Realization of the Living.* In: Cohen, R. S. and Wartofsky, M. W. (eds.): Boston Studies in the Philosophy of Science, Vol. 42, D. Riedel Publishing Co., Dordecht (1980).

[59] Maturana, H. and Varela, F.: *The Tree of Knowledge: The Biological Roots of Human Understanding.* Shambhala, Boston (1987).

[60] Maturana, H.: *The Organization of the Living: A Theory of the Living Organization.* In International Journal of Man-Machine Studies, Vol. 7 (1975), pp. 313-332.

[61] Mertins, K., Schallock, B. and Arlt, R. : *Ein Domänenmodell für das betriebliche Wissensmanagement.* In: ZwF 89, Carl Hanser Verlag, Munich (1994), pp. 512-513.

[62] Mingers, J.: *Self-Producing Systems: Implications and Applications of Autopoiesis.* Plenum Publishing, New York (1994).

[63] Morales, J. : *The Definition of Life.* http://home1.gte.net/lindanic/9801/life.htm (7/15/01). Verizon Communications Inc. (2001).

[64] Nickols, F.: *The Knowledge in Knowledge Management.* In: The Knowledge Management Yearbook 2000-2001, Butterworth Heinemann, Boston (2000), pp. 12-21.

[65] Nonoka, I.: *The Knowledge-Creating Company.* Harvard Business Review, November-December (1991), pp. 96-104.

[66] Nonoka, I. and Takeuchi, H.: *The Knowledge-Creating Company: How Japanese Companies Create the Dynamics of Innovation.* Oxford University Press (1995).

[67] Ocasio, W.: *How Do Organizations Think?* In: Lant, T. and Shapira, Z.: Organizational Cognition, Computation and Interpretation, LEA Publishers, Mahwah (2001), pp. 39-60.

[68] O'Haer, G.M P. and Chisholm, A.W.J. : *Disrtibuted artificial intelligence; an invaluable technique for the development of intelligent manufacturing systems*. Annals of the CIRP, Vol 39/1 (1990), pp 485-488.

[69] Orgel, L.: *Selection in vitro*. Proc. Royal Soc. of London (1979).

[70] Piaget, J.: *Meine Theorie der geistigen Entwicklung*. Fischer, Frankfurt am Main (1983).

[71] Piaget, J.: *The psychology of intelligence*. Littlefield Adams, Totowa. NJ (1972).

[72] Platon: *Theaitetos*. In: Sämtliche Werke. Band 4. Nach der Übersetzung von F. Schleiermacher herausgg. von W. F. Otto, E. Grassi und G. Plamböck, Rowohlt Taschenbuch Verlag, Hamburg (1958).

[73] Popper, K.: *Conjectures and Refutations: The Growth of Scientific Knowledge*. Routledge, London (1963).

[74] Popper, K.: *Evolutionary Epistemology*. in Pollard, J. W. (ed.): 'Evolutionary Theory: Paths into the Future.', John Wiley & Sons Ltd., London (1984).

[75] Popper, K.: *Objective Knowledge: An Evolutionary Approach*. The Clarendon Press, Oxford (1972).

[76] Pritschow, G. and Wiendahl, H.-P.: *Application of Control Theory for Production Logistics – Results of a Joint Project*. Annals of the CIRP, Vol 44/1 (1995), pp 421-424.

[77] Probst, G., Raub, S. and Romhardt K.: *Wissen managen: wie Unternehmen ihre wertvollste Ressource optimal nutzen*. Gabler Verlag, Frankfurt am Main, 3rd edition (1999)

[78] Probst, G. and Romhardt K.: *Bausteine des Wissensmanagements – ein praxisorientierter Ansatz*. http://www.cck.uni-kl.de/wmk/papers/public/Bausteine/bausteine.pdf (7/8/01) Arbeitsgemeinschaft Wissensmanagement, Universität Kaiserslautern, (1997)

[79] Radler, J.: *A Cognitive Perspective of Human Intelligence*. http://www.fau.edu/divdept/cla/honor/HTML/Project1.html (08/09/01), Florida Atlantic University, Florida (1999).

[80] Rao, A. S. and Georgeff, M. P.: *BDI agents: from theory to practice*. In: Proceedings of the First International Conference on Multi-Agent Systems, San Francisco (1995).

[81] Reese-Schaefer, W.: *Luhmann zur Einführung*. Julius Verlag, Hamburg (1992).

[82] Robb, F.: *Accounting – A Virtual Autopoietic System?* Systems Practice, Vol. 4, no. 3, (1991), pp. 215-235.

[83] Roberts-Witt, S. L.: *Practical Taxonomies*. In: The Knowledge Management Yearbook 2000-2001, Butterworth Heinemann, Boston (2000), pp. 214-222.

[84] Romhardt, K.: *Interventionen in die organisatorische Wissensbasis zwischen Theorie und Praxis – Welchen Beitrag kann die Systemtheorie leisten?* Cahier de recherche, HEC, Université de Genève, Geneva (1996).

[85] Roos, J. et al.: *Intellectual Capital: Navigating in the New Business Landscape*. New York University Press, New York (1998).

[86] Rorty, R.: *Consequences of Pragmatism.* University of Minnesota Press (1982).

[87] Rukeyser, M.: *Williard Gibbs.* Doubleday, New York (1942).
Reprint: Ox Bow Press, Woodbridge (1988)

[88] Russel, B.: *Human Knowledge.* Simon and Schuster, New York (1948).

[89] Sartwell, C.: *Why Knowledge is Merely True Belief.* The Journal of Philosophy 89, (1992), pp. 167-180.

[90] Saussure, F.: *Cours de linguistique générale.* By Tullio De Mauro, Payot, Paris (1916/1975).

[91] Schüppel, J.: *Wissensmanagement – Organisatorisches Lernen im Spannungsfeld von Wissens- und Lernbarrieren.* Gabler, Wiesbaden (1996).

[92] Schuster, P. and Sigmund, K. : *Vom Makromolekül zur primitiven Zelle – Das Prinzip der frühen Evolution.* In: Biophysik, Springer Verlag, Berlin (1982), pp 907-947.

[93] Silbiger, S. A.: *The 10-Day MBA.* Piatkus, London (1999).

[94] Sitkin, S. B.: *Commentary: The Theoretical Foundations of Organizational Cognition.* In: Lant, T. and Shapira, Z.: Organizational Cognition, Computation and Interpretation, LEA Publishers, Mahwah (2001), pp. 73-79.

[95] Srikantaiah, T. K.: *Knowledge Management: A Faceted Overview.* In: Srikantaiah, T. K., Koenig M. E.D.: Knowledge Management for the Information Professional, ASIS, Medford (1999), pp. 7-17.

[96] Stehr, N. : *Arbeit, Eigentum und Wissen: Zur Theorie von Wissensgesellschaften.* Suhrkamp, Frankfurt (1994).

[97] Sternberg, R. J.: *The Triachic Mind: A New Theory of Human Intelligence.* Penguin Books, New York (1988)

[98] Stewart, T. A. : *Brainpower.* Fortune: 123 (1991a), pp 44-60.

[99] Stewart, T. A. : *Getting Real About Brainpower.* Fortune: 132 (1995), pp. 201-203.

[100] Stewart, T. A. : *Intellectual Capital: Ten Years Later, How Far We've Come.* Fortune: 143:11 (2001).

[101] Stewart, T. A. : *Intellectual Capital: The New Wealth of Organizations.* Doubleday, New York (1997).

[102] Stewart, T. A. : *Now Capital Means Brains, Not Just Bucks.* Fortune: 123 (1991b), pp. 31-32.

[103] Stewart, T. A. : *Your Company's Most Valuable Asset: Intellectual Capital.* Fortune: 130 (1994), pp.68-74.

[104] Sveiby, K. E. and Mellander, K.: *Tango: Business from Knowledge Learning Guide.* Celemi, Malmoe (1994).

[105] Sveiby, K. E.: *The New Organizational Wealth: Managing and measuring intangible assets.* Berrett-Koehler, San Francisco (1997).

[106] Taylor, F. W.: *The Principles of Scientific Management*. Harper and Brothers Publishing, New York (1911).

[107] Teubner, G. (ed.): *Autopoietic Law: A New Approach to Law and Society*. Walter de Gruyter, Berlin / New York (1988).

[108] Teubner, G. and Febbrajo, A. (eds.): *State, Law and Economy as Autopoietic Systems: Regulation and Autonomy in a New Perspective*. European Yearbook in the Sociology of Law, Giuffre, Milan (1992).

[109] Thannhuber, M., Tseng, M. M., Bullinger H.-J. : *An Autopoietic Approach for Building Knowledge Management Systems in Manufacturing Enterprises*. The Annals of the CIRP Vol50/1 (2001), pp. 313-318.

[110] Torgersson, O.: *A Note on Declarative Programming Paradigms and the Future of Definitional Programming*. http://www.md.chalmers.se/pub/users/oloft/Papers/wm96/wm96.html (08/09/01) Department of Computing Science, Chalmers University of Technology and Göteborg University, Göteborg (1996).

[111] Tseng, M. M., Chen, Y. J., Gupter, A., Wang, L.: *Assembly Process Planning of Mass Customized Products Based on Living Systems Theory*. Submitted to The 35th CIRP International 3.Seminar on Manufacturing Systems, Hong Kong (2002).

[112] Varela, F. : *Principles of Biological Autonomy*. Elsevier, New York (1979).

[113] Wah, L.: *Behind the Buzz: The Substance of Knowledge Management*. In: The Knowledge Management Yearbook 2000-2001, Butterworth Heinemann, Boston (2000), pp. 307-317.

[114] Wah, L.: *Making Knowledge Stick*. In: The Knowledge Management Yearbook 2000-2001, Butterworth Heinemann, Boston (2000), pp. 145-156.

[115] Weick, K. M.: *The social psychology of organizing*. Random House, New York, 2nd edition (1979).

[116] Weinberg, R. A.: *Intelligence and IQ: Landmark Issues and Great Debates*. American Psychologist, v44n2, (1989), pp. 98-104.

[117] Whitaker, R.: *Autopoietic Theory: Deeper Discussion*. http://www.acm.org/siggroup/ois/auto/ATDefs.html (7/28/01), Association for Computing Machinery (1995a).

[118] Whitaker, R.: *Autopoietic Theory and Social Systems: Theory and Practice*. http://www.acm.org/siggroup/ois/auto/AT&Soc.html (12/15/01), Association for Computing Machinery (1995c).

[119] Whitaker, R.: *Self-Organization, Autopoiesis, and Enterprises*. http://www.acm.org/sigois/auto/Main.html (7/28/01), Association for Computing Machinery (1995b).

[120] Wittmann, W.: *Unternehmen und Unvollkommene Information: Unternehmerische Voraussicht – Ungewißheit und Planung*. Köln (1959).

[121] Wittmann, W.: *Wissen in der Produktion*. In: Handwörterbuch der Produktionswirtschaft, Stuttgart (1979), pp. 2261-2272.

[122] Wright, T. P.: *Factors Affecting the Cost of Airplanes*. In: Journal of the Aeronautical Science (1936), pp. 122-128.

[123] Zeleny, M. and Hufford, K. D.: *The Application of Autopoiesis in Systems Analysis: Are Autopoietic Systems Also Social Systems?* In: International Journal of General Systems, Vol. 21 (1992), pp. 145-160.

Glossary

Adaptation. Is the minimization of irritations in the interactions with the environment, being a necessary requirement for effective and successful behavior.

Agent, Software~. Executable code or a program that runs and interacts independently in an effort to accomplish its goals; Usually being executed in parallel with many other agents (multi agent systems) or conventional software.

Autopoiesis. Is the ability of a system to generate its specific constitution – its components (structure) and their interplay (organization) – on its own.

Business Process. Any sequence of activities enacted by an organization in order to achieve 'value added' and realize profits.

Capability, the employees' ~. An employee's communication of an ability to fulfill a certain task. The interface between system 'employee' and system 'enterprise', enabling the communication of 'What to achieve'. It is an entity that encapsulates any desired set of actions which is carried out by the employee.

Cognition. Is the actual 'acting' or 'behaving' within the cognitive domain.

Cognitive Domain. Is given by all those interactions (a system can enter) for which the system can successfully assemble a meaningful process as response to the triggering information (event).

Constraints, systems processing ~. All the building blocks – implementation level procedures – the system has incorporated (enactments available for disposition) and to which the system's control mechanism has to constrain its selections.

Control. The act of selecting or commanding steps and activities which defines the directives or the "How" a system responds to a given stimulus.

Declarative Processing. (Behavior-defining) processing, under which the response behavior of a system is auto-assembled (autonomous control) on trigger uniquely for every given stimulus (providing a notion of "What to achieve" – the logic).

Degrees of Freedom. A measure of variability, expressing the number of options a system faces or has available at a given time t.

Efficient Size of Operation. The production volume (size of operation) at which the average costs per unit are minimal.

Entropic Decay. The thermodynamic decay of order which any open system (characterized by its order) is subject to.

Entropy. A thermodynamic quantity specifying the amount of disorder or randomness in a system.

Epistemology. The philosophic discipline that provides the scientific framework for the theories to describe and understand human knowledge. It investigates the origin, nature, methods and limits of human knowledge.

Evolutional Learning. Changes in physical abilities and behavioral patterns that emerged and developed over generations driven by the natural balance of heredity vs. variety generation and evolutional selection and extinction.

Explicit Knowledge. (Traditional Knowledge Management) Acts, methods, principles, techniques and the like, which the knowledge bearer is aware of and which can be articulated. Knowledge of a formal and systematic type.

Implicit Knowledge. (Traditional Knowledge Management) Behavioral patterns and schemes, internalized procedures and the like, which one applies and executes without explicit awareness and thus cannot be articulated at a given moment.

Intellectual Capital (IC). A fourth production factor next to land, labor and capital. Regarded as a company's intangible asset, dominantly accounting for innovation and 'value-added' achieved by applying knowledge.

Intelligence. A methodological framework for continuous adaptation and the incorporation of behavioral patterns, realizing virtual evolution of pattern fragments that are populated and depreciated, based on evaluation schemes that ensure that the overall development proceeds in the right direction.

Intelligent Enterprise. An Enterprise that realizes declarative processing, the development of suitable control procedures and successful process building blocks – all enabling it to adapt quickly and successfully to a competitive and continuously changing environment.

Intelligent Learning. Specific behavioral changes and new behavior patterns a system incorporates within its own lifetime as a consequence of its behavioral effectiveness.

Information. A set of events that perturb a system, forcing the selection of system states. Information is a typical trigger for a system to start processing.

Information Management (IM). The well-organized collection and preparation of information targeted to support managerial decision processes.

Information Technology (IT). A number of technological branches, including computer and communication technologies, that deal with information processing, storing retrieval, transmission, etc.

Interpenetration. The reciprocal dependency of humans and social systems. Both only exist due to the other even though both are only environment to the other.

Knowledge, philosophic ~. Justified true belief; in the scientific world: the body of facts, relations, teachings, etc. that have been proven by verification, with the 'fundamental truth' being the aim and guidance for all scientific efforts.

Knowledge, from a microscopic perspective. (System knowledge, microscopically) Is the sum of all those incorporated processes that coordinate and constrain the use of system-specific degrees of freedom gained through system integration.

Knowledge, from a macroscopic perspective. (System knowledge, macroscopically) The set of constraints and control that enable the instantiation of those processes that are accessible through the system's structural setup (-> Declarative Processing).

Knowledge Workers. Employees, who are addressed only with a notion of 'What to achieve' whereas they decide themselves how this is done – autonomously solving problems and seeking for solutions.

Logic.[49] Classically, the teaching of consistent and ordered thinking, thus provides the techniques to the deduction of a meaningful response. Strongly reduced by modern philosophy, it in this work represents the interface (set of stimuli) enabling a mediation between two systems. It is the techniques two systems adopt to post suitable stimuli in expectation of a specific (anticipatable) response behavior (posting system) and to respond to a given stimuli with a behavior that lets anticipate success (receiver system).

---

[49] Computer Science stipulated logic to be a highly complex philosophical concept to allow a new discussion inflamed in the 20th century: Artificial Intelligence (AI). Classically 'Logic' was defined as: "... the teaching of consistent and ordered thinking" (KUNZMANN, BURKHARD and WIEDMANN, 1998 [46], p. 13). Philosophers classically separate logic in two parts, the elementary teaching (discussing 'term', 'judgment' and 'conclusion') and the teaching of methods (discussing methods of investigation and proof). Logic was extensively discussed already by Aristotle (384-324b.C.). Logic in this sense provides the techniques for the deduction of a meaningful response. The deduction was thought to be based on explicit knowledge and was a sort of art the human mind was able to carry out. Biology, of course, later thought that deduction is nothing more than the involvement of incorporated response patterns built up empirically (empiric logic). As early as in the 19th Century the mathematician and philosopher Gottlob Frege (1848-1925) built what is known as 'symbolic logic' – formalizing and simplifying the classical concepts by using symbolic languages. He introduced the predicate calculus and quantifications. Most important, he assumed that every logical expression, sentence or formula has content and meaning (Ger.: 'Sinn'). Later, mathematicians and computer scientists adopted his work. The fundamental underlying assumption in their work was that 'Logic' could be used to make explicit and transfer 'Meaning' from one domain (e.g. humans) to another (e.g. computer systems). Symbolic logic, however, is a synthetic logic, which, as we all know today, is very different from empiric logic, which governs the behavior of biological systems and in particular humans – the poor results of 20 years of AI research are a direct consequence! Modern philosophy strongly reduced this once so blooming concept. In this work it merely represents the interface (in the form of a set of stimuli) enabling the mediation between two systems. It is the techniques two systems adopt to post suitable stimuli in expectation of a specific (anticipatable) response behavior (posting system) and to respond to a given stimuli with a behavior that lets anticipate success (receiver system). This in particular does not imply a transfer of meaning! The posting system will be governed by a system of meaning that is highly different from that of the receiver system and the stimulus does mean something very different to the sender than it does to the receiver. However, it allows their mediation in that it allows anticipating response or success based on empiric logic. By posting the stimuli the posting system logically expects a certain response behavior (meaningful response), whereas the stimulus provides a notion of "What to achieve" to the receiver system, which is the logic of 'What' it will enact (meaningful behavior). Example: A guest (posting system) in a restaurant, holds her empty tea cup towards the waiter (receiver system). The guest '**logically**' anticipates that the waiter fills the cup, as she indicates that she is thirsty (meaning). The waiter, perceiving the presented cup (stimulus), will indeed fill it. The stimulus provides him with the **logic** of what he does at the given moment: fill the cup. He does this as he anticipates that the guest is happy afterwards, his manager expects him to do this and he is tipped properly thereafter (meaning) – whether or not the guest is thirsty.

Logic, in Declarative Processing. The notion of 'What to achieve', the goal or desired result, which is implied by the situation-specific challenge that any given stimulus might be to the system.

Marginal Costs. The amount by which the total costs change if production output is increased or decreased by one unit.

Marginal Revenue. The additional revenue earned if production output is increased by one unit.

Market Power. The freedom a company has to set its own product price and with it control its sales volume without suffering immediate losses.

Meaning. The macroscopic manifestations of microscopic effects (laws and principles) that direct the transitions between system states. It is microscopically represented by the causal relationships and their consequences. Classically it was understood as the sense, significance or value something has to the observer.

Monopolistic Competition. A market in which many suppliers sell similar products and close substitutes but not identical goods.

Monopoly. A market structure in which one supplier is the sole seller of a product without the availability of close substitutes.

Net Welfare Loss (Deadweight Loss). The reduction of the total surplus – consumer surplus (=max. accepted price minus real price) plus producer surplus (=sales price minus costs).

Neuronal Network. A computer software implementing a network of input-processing elements (neurons) that attempts to imitate the way a biological brain works.

Neurotransmitter. Chemicals that transmit a stimulus across the junction (synapse) separating one nerve cell (neuron) from another nerve cell or a muscle.

Oligopoly. A market in which a specific product is produced by only a few bulk suppliers.

Ontology. Classically, the branch of metaphysics that studies the nature of existence or being as such. Here it is understood as a set of classes identified to exist and existential assumptions on them – e.g. 'Fact Names' (sensors) and possible 'Fact Values' (readings).

Organization, a system's ~. (Greek: organikos – serving as instruments, instrumental) Is the instrumental participation of the components in the constitution of the unit. It defines the components' functional interplay and all possible processes that the system can enact.

Organizational Learning (OL). A process of continuous transformation and improvement by a strong dedication of the organization to educate and train each and every member and facilitate her/his learning.

Participant, a system's ~. The major players involved at the formation of the system. The cells of plants or animals; the employees, managers, directors, etc. of the system enterprise; … (not to be confused with the system's structural elements!)

Perception (of Information). Is the successful assembly of a meaningful process in response to the triggering information.

Perfect Competition or Perfect Market. A hypothetical market with many buyers and many sellers trading identical products.

Process. A course of events and actions that are coupled by the principle of cause and effect (event-action-world). The sequence of states a system passes through with each transition following an underlying principle or guiding logic (State world).

Process Building Block (PBB). A generic unit, encapsulating a set of microscopic activities or implementation level procedures with a known enactment (executable on trigger), which are chained by the system's control mechanism to a macroscopic process. The constraints of the control process.

Processing. Is the sequential selection of actions, each forcing the system into new system states (process).

Process-Driven Management. A management paradigm which organizes and aligns all managerial efforts, targets and contents to a framework of processes that are established to define and run the enterprise.

Project-Driven Management. A management paradigm which organizes and aligns all managerial efforts, targets and contents to a framework of projects that are established to define and run any activity within the enterprise.

Rational Behavior. A behavior that is based on interactions that have a rational character, involving the use of explicit knowledge.

Scope, a system's ~. The range or set of structural components to which it can directly apply its control mechanism.

Selection. The macroscopic manifestations of microscopic processes that enact the transitions between microscopic states.

Stimulus. An event (figuratively also: situation, condition, signal, etc.) that excites an organism/system and may provoke a response behavior.

Structure, a system's ~. (Latin: structura – a fitting together, building; Latin: structum – something that has been built) is the set of physical components in a given space. It determines the space in which the system exists and can be perturbed.

System, a classic ~. Any part of the material universe which is separated in thought from the rest for the purpose of considering and discussing the various changes that may occur within it under various conditions.

System, a real ~. A set of processes, structural components and their interactions (organization) that define an entity of its own, realizing a character which on one hand can only be achieved in this entity and on the other hand is existential to the entity.

System Level. A classification of systems based on the type of participants they integrate. The higher the 'System Level' of a system, the larger the number of systems on lower 'System Levels' that have been nested in order to integrate it.

Tacit Knowledge. In this work used as a synonym for 'Implicit Knowledge'. See Implicit Knowledge.

Value Added. The price the customer is willing to pay for all operational steps required to develop, form and prepare the product (in all its features) and its environment (distribution channel, image, services, etc.).

Workflow. The sequence of jobs/actions that forms as a result of the enactment of an enterprise process.

Index

Printing and Binding: Strauss GmbH, Mörlenbach

GPSR Compliance

The European Union's (EU) General Product Safety Regulation (GPSR) is a set of rules that requires consumer products to be safe and our obligations to ensure this.

If you have any concerns about our products, you can contact us on ProductSafety@springernature.com

In case Publisher is established outside the EU, the EU authorized representative is:

Springer Nature Customer Service Center GmbH
Europaplatz 3
69115 Heidelberg, Germany

Batch number: 10725973

Printed by Printforce, the Netherlands